Current ed. 04/03 JM.

SHIPS' CODE AND DECODE BOOK

Incorporating the International Meteorological
Codes for Weather Reports from and to Ships and
the Analysis Code for use of Shipping

LONDON: The Stationery Office

Decimal Index
551.509.1

10 752 2796

© Crown Copyright 1996
Published for the Meteorological Office under licence from the Controller of Her Majesty's Stationery Office
Applications for reproduction to be made in writing to the Copyright Unit, Her Majesty's Stationery Office, St Clements House, 2–16 Colegate, Norwich NR3 1BQ
First published 1948
Twelfth edition 1996
2nd Impression 1997

ISBN 0 11 400368 8

CONTENTS

Page

Introduction 2

Section 1   Code Forms (FM 13-X etc.), and instructions regarding omission of information in the coded groups 3

Section 2   The individual code groups 4

Section 3   Code tables (Tables 1 to 27) 8

Section 4   The Atlantic Weather Bulletin 31

Section 5   The International Analysis Code I.A.C. FLEET (FM 46-IV), including appropriate code tables (Tables 28 to 36) 36

Section 6   Maritime Forecast Code ("MAFOR") FM 61-IV, including appropriate code tables (Tables 37 to 45) 47

Conversion table, feet/metres 53

Symbols used on radio-facsimile charts 54

Alphabetical list of code symbols 56

Index 58

Plotting Chart for use with Atlantic Weather Bulletin for Shipping 60

Port Meteorological Officers, Great Britain 61

INTRODUCTION

The codes in this book (FM12-X Ext., FM13-X, etc.) came into force at 0000 UTC on 1st January 1982, with amendments made to the codes on 1st November 1987 and 2nd November 1994. These codes are to be used for all coded weather messages from British ships.

This book is intended for both coding and decoding. For the coding of reports, full details are given in Sections 1–3. Sections 4–6 give details of the Atlantic Weather Bulletin and other international codes for shipping.

Notes appear below the relevant group in Section 2 where a whole code group is concerned. A note is also given in Section 2 where an individual code letter not having a corresponding code table in Section 3 is concerned. In cases where an individual code letter is concerned which does have a code table in Section 3 the notes are appended to the code table itself.

The notes appended to the code groups and code tables in this book are more detailed than in the Ships' Code Card because the code card is produced primarily for quick ready use when making and coding an observation.

Instructions are given on page 3 regarding the procedure to be adopted when coding a report in which not all the information is available. They should be read through carefully.

SECTION 1
THE CODING OF SHIPS' WEATHER REPORTS

FM13-X — Selected Ship's Code
BBXX D .... D YYGGi_w 99$L_aL_aL_a$ $Q_cL_oL_oL_oL_o$ $i_{Ri}i_xhVV$ Nddff (00fff)
1s_nTTT 2$s_nT_dT_dT_d$ (3$P_oP_oP_oP_o$) 4PPPP 5appp (6RRRt_R) 7wwW_1W_2
8$N_hC_LC_MC_H$ (9hh///) 222D_sv_s 0$s_sT_wT_wT_w$ (1$P_{wa}P_{wa}H_{wa}H_{wa}$)
2$P_wP_wH_wH_w$ 3$d_{w1}d_{w1}d_{w2}d_{w2}$ 4$P_{w1}P_{w1}H_{w1}H_{w1}$ 5$P_{w2}P_{w2}H_{w2}H_{w2}$ 6$I_sE_sE_sR_s$
70$H_{wa}H_{wa}H_{wa}$ 8$s_wT_bT_bT_b$ ICE $c_iS_ib_iD_iz_i$

Supplementary Ship's Code
BBXX D .... D YYGGi_w 99$L_aL_aL_a$ $Q_cL_oL_oL_oL_o$ $i_{Ri}i_xhVV$ Nddff (00fff)
1s_nTTT 4PPPP 7wwW_1W_2 8$N_hC_LC_MC_H$ 222D_sv_s 6$I_sE_sE_sR_s$
ICE $c_iS_ib_iD_iz_i$

Auxiliary Ship's Code
BBXX D .... D YYGGi_w 99$L_aL_aL_a$ $Q_cL_oL_oL_oL_o$ $i_{Ri}i_x$/VV Nddff (00fff)
1s_nTT/ 4PPP/ 7wwW_1W_2 222D_sv_s 6$I_sE_sE_sR_s$ ICE $c_iS_ib_iD_iz_i$

Offshore Observer's Code
BBXX $A_1b_wn_bn_bn_b$ YYGGi_w 99$L_aL_aL_a$ $Q_cL_oL_oL_oL_o$ $i_{Ri}i_xhVV$ Nddff
(00fff) 1s_nTTT 2$s_nT_dT_dT_d$ 4PPPP 5appp 7wwW_1W_2 8$N_hC_LC_MC_H$
222D_sv_s 0$s_sT_wT_wT_w$ 1$P_{wa}P_{wa}H_{wa}H_{wa}$ 2$P_wP_wH_wH_w$ 3$d_{w1}d_{w1}d_{w2}d_{w2}$
4$P_{w1}P_{w1}H_{w1}H_{w1}$ 5$P_{w2}P_{w2}H_{w2}H_{w2}$ 6$I_sE_sE_sR_s$ 70$H_{wa}H_{wa}H_{wa}$ 8$s_wT_bT_bT_b$
ICE $c_iS_ib_iD_iz_i$

Mandatory reporting of information in the coded groups
The following groups must always be reported:

1. Call sign or identification number followed by the next five groups.

2. Group 222D_sv_s.

3. Groups 7wwW_1W_2, 8$N_hC_LC_MC_H$, 2$P_wP_wH_wH_w$, 3$d_{w1}d_{w1}d_{w2}d_{w2}$ and 4$P_{w1}P_{w1}H_{w1}H_{w1}$, i.e. weather, clouds, waves and swell.

4. Supplementary Ships which have been supplied with sea temperature equipment, should report the groups 222D_sV_s and, 0$s_sT_wT_wT_w$.

Omission of information in coded groups

1. If information for part of a coded group is not available, a solidus (/), or the appropriate number of these must be reported instead.*

2. If information is not available for groups with indicators, they should be omitted entirely.

3. Groups 00fff, 3$P_oP_oP_oP_o$, 6RRRt_R, 9hh///, 1$P_{wa}P_{wa}H_{wa}H_{wa}$ and 70$H_{wa}H_{wa}H_{wa}$ are not routinely reported by ships.

---

* To avoid confusion with the figure 1, the solidus should be crossed through with a small horizontal line, both in the log and in the radio message, thus ⧸.

SECTION 2
THE INDIVIDUAL CODE GROUPS

Group D . . . . D

Ship's call sign

Group $A_1B_wn_bn_bn_b$

A_1 WMO Region in which offshore platform has been deployed.
b_w Sub-area of A_1.
$n_bn_bn_b$ Serial number of platform.

Group $YYGGi_w$

YY   Day of the month (UTC).
GG   Time of observation to nearest hour (UTC).
i_w Wind speed indicator (Table 1).

$99L_aL_aL_a\ Q_cL_oL_oL_oL_o$

99   Indicator figures for SHIP report.
$L_aL_aL_a$ Latitude in degrees and tenths.
Q_c Quadrant of the globe (Table 2).
$L_oL_oL_oL_o$ Longitude in degrees and tenths.

Note. The tenths figure of latitude and longitude is obtained by dividing the number of minutes by 6, disregarding the remainder.

Group i_Ri_xhVV

i_R Indicator figure for precipitation group (Table 3).
i_x Indicator for type of station and for present and past weather phenomena (Table 4).
h   Height of base of the lowest cloud in the sky (irrespective of type). (Table 5). See notes under $8N_hC_LC_MC_H$.
VV   Horizontal visibility at surface (Table 6).

Group Nddff

N   Total amount of cloud in eighths (oktas) (Table 7).
dd   True direction, in tens of degrees, from which the surface wind is blowing (Table 8).
ff   Speed of surface wind in knots. This value is derived from the observed Beaufort Force (Table 9).

Group 00fff

00   Indicator figures for this group.
fff   Wind speed in units indicated by i_w of 99 units or more.

Notes. (1) U.K. Observing ships are not issued with anemometers. Wind speed in these ships should always be estimated from the appearance of the sea.

(2) When wind speed is 99 units or more ff is reported as 99 and the group 00fff is reported immediately after group Nddff.

Group $1s_nTTT$

1	Indicator figure for this group.
s_n	Sign of temperature (Table 10).
TTT	Air temperature in whole degrees and tenths.

Group $2s_nT_dT_dT_d$

2	Indicator figure for this group.
s_n	Sign of temperature (Table 10).
$T_dT_dT_d$	Dew-point temperature in whole degrees and tenths.

See Met.0.938, Dew-point tables (An abbreviated Table appears between pages 27–30).

Group 4PPPP

4	Indicator figure for this group.
PPPP	Pressure in millibars and tenths (omitting thousands figure) e.g. 998.2mb coded as 9982; 1014.7mb as 0147.

Group 5appp

5	Indicator figure for this group.
a	Characteristic of barometric tendency during the three hours preceding the time of observation (Table 11).
ppp	Amount of barometric tendency (i.e. net change in barometer reading) during the three hours preceding the time of observation, expressed in tenths of a millibar.

Group $6RRRt_R$

This group is not intended for use by British Selected Ships but may be used by Ocean Weather Ships, or by ships of some other countries, when they are supplied with rain-gauges.

Group $7wwW_1W_2$

7	Indicator figure for this group.
ww	Present weather (Table 12).
W_1W_2	Past weather, i.e. weather since last synoptic hour (Table 13).

Group $8N_hC_LC_MC_H$

8	Indicator figure for this group.
N_h	Amount of all low cloud present or, if no low cloud present, all the medium cloud present (Table 7).
C_L	Type of low cloud (Sc, St, Cu, Cb) (Table 14).
C_M	Type of medium cloud (Ac, As, Ns) (Table 15).
C_H	Type of high cloud (Ci, Cc, Cs) (Table 16).

Notes. (1) If the sky is not discernible owing to fog or other surface phenomena, this group is coded as 89///. In group i_Ri_xhVV, h = /.

(2) If there is no cloud in the sky the group is coded as 80000 and h = 9.

(3) If there is fog, and the sky is discernible through the fog, the cloud type, height and amount are reported as if no fog were present.

(4) If there is more than one layer of low cloud, N_h refers to the total amount of the sky they cover and h refers to the lowest of these layers.

(5) If there is no low cloud present, N_h refers to the total amount of medium cloud. If there is neither low nor medium cloud N_h = 0, h = 9.

Group 222$D_s v_s$

222	Section Indicator.
D_s	Ship's course made good during the three hours preceding the time of observation (Table 17).
v_s	Ship's average speed made good during the three hours preceding the time of observation (Table 18).

Group 0$s_s T_w T_w T_w$

0	Indicator figure for this group.
s_s	Sign and type of sea surface temperature (Table 26).
$T_w T_w T_w$	Sea surface temperature in degrees and tenths.

Group 1$P_{wa}P_{wa}H_{wa}H_{wa}$

1	Indicator figure for this group.
$P_{wa}P_{wa}$	Period of waves (wave recorder) in seconds.
$H_{wa}H_{wa}$	Height of waves (wave recorder) in ½ metres.

Note. This group is only reported by vessels equipped with a Wave Recorder.

Group 2$P_w P_w H_w H_w$

2	Indicator figure for this group.
$P_w P_w$	Period of sea waves in seconds.
$H_w H_w$	Height of sea waves in units of ½ metres (i.e. 01 = 0.5 metre) (See p. 53).

Notes. (1) The group 2$P_w P_w H_w H_w$ refers to sea waves (wind waves) only and direction is not included because this is the same as wind direction.
 (2) If the estimation of period of sea waves is impossible owing to confused sea, $P_w P_w$ is reported as 99. Solidi (//) are used when the period is not observed for any other reason.
 (3) If there is a swell with no sea waves the sea wave group should be entered as 20000 in the logbook but omitted from the radio message.
 (4) The average value of the wave height (vertical distance between trough and crest) is reported, as obtained from the larger well-formed waves of the wave system being observed.

Groups 3$d_{w1}d_{w1}d_{w2}d_{w2}$, 4$P_{w1}P_{w1}H_{w1}H_{w1}$, 5$P_{w2}P_{w2}H_{w2}H_{w2}$

3	Indicator figure for this group.
$d_{w1}d_{w1}$	Direction, in tens of degrees, from which the first swell waves are coming (Table 8).
$d_{w2}d_{w2}$	Direction, in tens of degrees from which the second swell waves are coming (Table 8).
4	Indicator figure for this group.
$P_{w1}P_{w1}$	Period of first swell waves in seconds.
$H_{w1}H_{w1}$	Height of first swell waves in units of ½ metres. (See p. 53).
5	Indicator figure for this group.
$P_{w2}P_{w2}$	Period of second swell waves in seconds.
$H_{w2}H_{w2}$	Height of second swell waves in units of ½ metres. (See p. 53).

Notes. (1) When it is possible to make a clear distinction between sea and swell, one or both swell groups should be reported.
 (2) If only one swell group is reported $d_{w2}d_{w2}$ is reported as // and group 5$P_{w2}P_{w2}H_{w2}H_{w2}$ is omitted.

Group $6I_sE_sE_sR_s$

6	Indicator figure for this group.
I_s	Type of ice accretion (Table 19).
E_sE_s	Thickness of ice accretion in cm.
R_s	Rate of ice accretion (Table 20).

Note. This group can be reported in plain language preceded by the word ICING.

Group $70H_{wa}H_{wa}H_{wa}$

70	Indicator figures for this group.
$H_{wa}H_{wa}H_{wa}$	Wave height measured in units of 0.1 metre.

Note. This group is reported by sea stations that have the capability of measuring wave height in units of 0.1 metre.

Group $8s_wT_bT_bT_b$

8	Indicator figure for this group.
s_w	Indicator for the sign and type of wet bulb temperature reported (Table 27).
$T_bT_bT_b$	Wet bulb temperature in degrees and tenths.

Group ICE $c_iS_ib_iD_iz_i$

c_i	Concentration or arrangement of sea ice (Table 21).
S_i	Stage of development (Table 22).
b_i	Ice of land origin (Table 23).
D_i	Bearing of ice edge (Table 24).
z_i	Ice situation and trend of conditions over the preceding 3 hours (Table 25).

Notes. (1) The group must always be preceded by the word ICE.
(2) The group should be reported whenever sea ice and/or ice of land origin is observed.
(3) The group can be reported in plain language preceded by the word ICE.

SECTION 3
CODE TABLES 1–25

Table 1

i_w = Wind speed indicator

Code figure
- 0   Wind speed estimated   ⎫
- 1   Wind speed from anemometer   ⎬ speed in metres/sec.
- 3   Wind speed estimated   ⎫
- 4   Wind speed from anemometer   ⎬ speed in knots

Table 2

Q_c = Quadrant of the globe

Code figure	Latitude	Longitude
1	N	E
3	S	E
5	S	W
7	N	W

Note. On the equator and 0° and 180° meridians the choice of Q_c values is left to the observer.

Table 3

i_R = Precipitation indicator

Code figure
- 1   Group included.
- 2   Group included in section 3 (not normally reported by ships).
- 3   Group omitted — No precipitation.
- 4   Group omitted — Precipitation amount not available.

Table 4

i_x = Type of station and present and past weather indicator

Code figure
- 1   Manned station, $7wwW_1W_2$ included.
- 2   Manned station, $7wwW_1W_2$ omitted (no significant phenomenon).
- 3   Manned station, $7wwW_1W_2$ omitted (no data available).
- 4   Automatic station, $7wwW_1W_2$ included.
- 5   Automatic station, $7wwW_1W_2$ omitted (no significant phenomenon).
- 6   Automatic station, $7wwW_1W_2$ omitted (no data available).
- 7   Automatic station, $7w_aw_a//$ included.

Table 5

h = Height of base of lowest cloud

Code figure	Feet	Metres
0	0 to 150	0 to 50
1	150 to 300	50 to 100
2	300 to 600	100 to 200
3	600 to 1,000	200 to 300
4	1,000 to 2,000	300 to 600
5	2,000 to 3,000	600 to 1,000
6	3,000 to 5,000	1,000 to 1,500
7	5,000 to 6,500	1,500 to 2,000
8	6,500 to 8,000	2,000 to 2,500
9	8,000 or more or no cloud	2,500 or more or no cloud
/	Height of base of cloud unknown	

Notes. (1) If the sky is not discernible owing to fog or other surface phenomena the base of the cloud is reported as /.

(2) A height exactly equal to one of the heights in the table is reported by the higher code figure, e.g. a height of 2,000 feet should be reported by code figure 5.

(3) It should be noted that N_h, C_L and h are not necessarily directly related to each other, for example:

    (i) 2/8 St. at 500 ft.
        3/8 Sc. at 1,500 ft.
        4/8 Sc. at 3,500 ft.
        Total amount of low cloud 7/8: coded as 875// and h is coded as 2.
    (ii) 3/8 St. at 800 ft.
        2/8 Cb. at 2,500 ft.
        2/8 Sc. at 4,000 ft.
        Total amount of low cloud 6/8: coded as 869// and h is coded as 3.
    (iii) 1/8 St. at 700 ft.
        5/8 St. at 1,000 ft.
        8/8 Sc. at 4,000 ft.
        Total amount of low cloud 8/8: coded as 885// and h is coded as 3.

Table 6

VV = Horizontal visibility at surface

Code figure	Distance km	yards	Code figure	Distance km	yards	Code figure	Distance km	yards	Code figure	Distance km	n. miles	Code figure	Distance km	n. miles
00	< 0.0	< 110	20	2.0	2,118	40	4.0	4,376	60	10	5.4	80	30	16.2
01	0.1	110	21	2.1	2,297	41	4.1	4,485	61	11	5.9	81	35	18.9
02	0.2	220	22	2.2	2,406	42	4.2	4,594	62	12	6.5	82	40	21.6
03	0.3	330	23	2.3	2,516	43	4.3	4,737	63	13	7.0	83	45	24.3
04	0.4	440	24	2.4	2,625	44	4.4	4,813	64	14	7.6	84	50	27.0
05	0.5	550	25	2.5	2,735	45	4.5	4,923	65	15	8.1	85	55	29.7
06	0.6	660	26	2.6	2,844	46	4.6	5,032	66	16	8.6	86	60	32.4
07	0.7	770	27	2.7	2,953	47	4.7	5,141	67	17	9.2	87	65	35.1
08	0.8	880	28	2.8	3,063	48	4.89	5,251	68	18	9.7	88	70	37.8
09	0.9	990	29	2.9	3,172	49	4.9	5,360	69	19	10.3	89	> 70	> 37.8
10	1.0	1,000	30	3.0	3,282	50	5.0	5,470	70	20	10.8	**90**	**< 0.05**	**< 0.03**
11	1.1	1,210	31	3.1	3,391	51			71	21	11.3	**91**	**0.05**	**0.03**
12	1.2	1,313	32	3.2	3,500	52	Not used		72	22	11.9	**92**	**0.2**	**0.1**
13	1.3	1,422	33	3.3	3,610	53			73	23	12.4	**93**	**0.5**	**0.3**
14	1.4	1,532	34	3.4	3,719	54		n. miles	74	24	13.0	**94**	**1**	**0.5**
15	1.5	1,641	35	3.5	3,829	55			75	25	13.5	**95**	**2**	**1.1**
16	1.6	1,750	36	3.6	3,938	56	6	3.2	76	26	14.0	**96**	**4**	**2.2**
17	1.7	1,859	37	3.7	4,047	57	7	3.8	77	27	14.6	**97**	**10**	**5.4**
18	1.8	1,969	38	3.8	4,157	58	8	4.3	78	28	15.1	**98**	**20**	**11**
19	1.9	2,075	39	3.9	4,266	59	9	4.9	79	29	15.7	**99**	**≥ 50**	**≥ 27**
													(See note (1) below)	

The symbol < indicates "less than". The symbol > indicates "more than". The symbol ≥ indicates "more than or equal to".

Notes. — (1) **The 90–99 decade is always employed in ship reports** for the reason that horizontal visibility cannot be determined with greater accuracy at sea. The full table is only shown here to enable reports from land stations to be decoded aboard ships.

(2) If the horizontal visibility is not the same in different directions, the shorter distance is coded.

(3) If the observed horizontal visibility is between two of the distances given in the table, the code figure for the shorter distance is reported, e.g. if the distance is estimated to be between 0.3 and 0.5 n.miles the code figure 93 is reported.

(4) In the international scale the distances for all code figures are expressed in metres. The visibilities listed above are the equivalent distances in nautical miles.

Table 7

N = Total amount of cloud
N_h = Amount of sky covered by all low cloud (or medium if no low cloud is present)

Code figure	
0	None
1	1 eighth of sky covered or less, but not zero.
2	2 eighths of sky covered.
3	3 eighths of sky covered.
4	4 eighths of sky covered.
5	5 eighths of sky covered.
6	6 eighths of sky covered.
7	7 eighths of sky covered or more, but not 8 eighths.
8	8 eighths (sky completely covered).
9	Sky obscured by fog or other meteorological phenomena.
/	Cloud cover obscured for other reasons or not observed.

Notes. (1) A "trace" would be coded under code figure 1, which should be used for amounts up to ⅛.
(2) "Overcast but with openings" would be coded under code figure 7.
(3) N, N_h is reported as 0 when blue sky or stars are seen through existing fog without any trace of cloud being visible.

Table 8

dd = Direction from which the surface wind is blowing
$d_{w1}d_{w1}$, $d_{w2}d_{w2}$ = Direction from which swell waves are coming

Code figures	True direction (degrees)	Code figures	True direction (degrees)
00*	Calm (no waves)	19	190
01	010	20	200
02	020	21	210
03	030	22	220
04	040	23	230
05	050	24	240
06	060	25	250
07	070	26	260
08	080	27	270
09	090	28	280
10	100	29	290
11	110	30	300
12	120	31	310
13	130	32	320
14	140	33	330
15	150	34	340
16	160	35	350
17	170	36	360
18	180		

* 00 is used only for calm (or no waves) and not for wind or wave direction North which is coded as 36.

Notes. (1) When the wind or wave direction is indeterminate code figure 99 should be reported. This figure will not normally be used if the wind speed is over 5 knots.
(2) When there is no wind, ddff is coded as 0000.
(3) In the case of half-way values (both wind and waves), the higher ten-degree value is coded, e.g. 125° is coded as 13.

Table 9

Beaufort Scale of Wind Force

Beaufort scale number	Mean wind speed		Limits of wind speed			Descriptive terms	Sea criterion (See photographs in *State of Sea* booklet or in *Marine Observer's Handbook*)	Probable height of waves in metres*	Probable maximum height of waves in metres*
	knots	metres per second	knots	metres per second					
			Measured at a height of 10 metres above sea level						
0	00	0.0	Less than 1	0.0–0.2		Calm	Sea like mirror.	—	—
1	02	0.8	1–3	0.3–1.5		Light air	Ripples with the appearance of scales are formed but without foam crests.	0.1	0.1
2	05	2.4	4–6	1.6–3.3		Light breeze	Small wavelets, still short but more pronounced; crests have a glassy appearance and do not break.	0.2	0.3
3	09	4.3	7–10	3.4–5.4		Gentle breeze	Large wavelets. Crests begin to break. Foam of glassy appearance. Perhaps scattered white horses.	0.6	1.0
4	13	6.7	11–16	5.5–7.9		Moderate breeze	Small waves, becoming longer; fairly frequent white horses.	1.0	1.5
5	19	9.3	17–21	8.0–10.7		Fresh breeze	Moderate waves, taking a more pronounced long form; many white horses are formed. (Chance of some spray.)	2.0	2.5
6	24	12.3	22–27	10.8–13.8		Strong breeze	Large waves begin to form; the white foam crests are more extensive everywhere. (Probably some spray.)	3.0	4.0
7	30	15.5	28–33	13.9–17.1		Near gale	Sea heaps up and white foam from breaking waves begins to be blown in streaks along the direction of the wind.	4.0	5.5

Beaufort Scale of Wind Force —continued

Beaufort scale number	Mean wind speed		Limits of wind speed			Descriptive terms	Sea criterion (See photographs in *State of Sea* booklet or in *Marine Observer's Handbook*)	Probable height of waves in metres*	Probable maximum height of waves in metres*
	knots	metres per second	knots	metres per second					
	Measured at a height of 10 metres above sea level								
8	37	18.9	34–40	17.2–20.7		Gale	Moderately high waves of greater length; edges of crests begin to break into spindrift. The foam is blown in well-marked streaks along the direction of the wind.	5.5	7.5
9	44	22.6	41–47	20.8–24.4		Strong gale	High waves. Dense streaks of foam along the direction of the wind. Crests of waves begin to topple, tumble and roll over. Spray may affect visibility.	7.0	10.0
10	52	26.4	48–55	24.5–28.4		Storm	Very high waves with long overhanging crests. The resulting foam in great patches is blown in dense white streaks along the direction of the wind. On the whole the surface of the sea takes on a white appearance. The tumbling of the sea becomes heavy and shocklike. Visibility affected.	9.0	12.5
11	60	30.5	56–63	28.5–32.6		Violent storm	Exceptionally high waves. (Small and medium-sized ships might be for a time lost to view behind the waves.) The sea is completely covered with long white patches of foam lying along the direction of the wind. Everywhere the edges of the wave crests are blown into froth. Visibility affected.	11.5	16.0
12	—	—	64 and over	32.7 and over		Hurricane	The air is filled with foam and spray. Sea completely white with driving spray; visibility very seriously affected.	14 and over	—

* These columns are added as a guide to show roughly what may be expected in the open sea, remote from land. In enclosed waters, or when near land with an off-shore wind, wave heights will be smaller and the waves steeper.

Notes: — (a) It must be realized that it will be difficult at night to estimate wind force by the sea criterion.
(b) The lag effect between the wind getting up and the sea increasing should be borne in mind.
(c) Fetch, depth, swell, heavy rain and tide effects should be considered when estimating the wind force from the appearance of the sea.

Table 10

s_n = Sign of air temperature (not for sea temperature — see Table 26)

Code figure
- 0   Temperature positive or zero.
- 1   Temperature negative.

$$T_d T_d T_d = \text{Dew-point}$$

An abbreviated dew-point table for screen readings will be found on pages 27–30. Extended dew-point tables are published separately in Met.O.938.

Table 11

a = Characteristic of barometric tendency during the three hours preceding the time of observation

Code figure	Trace	Description of curve	Pressure *now*, compared with 3 hours ago
0		Rising, then falling Rising, then falling	The same Higher
1		Rising, then steady Rising, then rising more slowly	Higher
2		Rising (steadily or unsteadily)	Higher
3		Falling, then rising Steady, then rising Rising, then rising more quickly	Higher
4		Steady	The same
5		Falling, then rising Falling, then rising	The same Lower
6		Falling, then steady Falling, then falling more slowly	Lower
7		Falling (steadily or unsteadily)	Lower
8		Steady, then falling Rising, then falling Falling, then falling quickly	Lower

Table 12

ww = Present weather

ww = 00–49.   NO PRECIPITATION AT THE STATION* AT THE TIME OF OBSERVATION

ww = 00–19.   No precipitation, fog, duststorm, sandstorm drifting or blowing snow at the station* at the time of observation or during the preceding hour, except for 09 and 17.

No phenomena except clouds	00	Cloud development not observed or not observable.	Characteristic of the state of sky during the past hour.
	01	Clouds generally dissolving or becoming less developed.	
	02	State of sky on the whole unchanged.	
	03	Clouds generally forming or developing.	
Haze, dust, sand or smoke	04	Visibility reduced by smoke, e.g. veldt or forest fires, industrial smoke or volcanic ashes.	
	05	Haze. †	
	06	Widespread dust in suspension in the air, not raised by wind at or near the station at the time of observation.	
	07	Dust or sand raised by wind at or near the station at the time of observation, but no well-developed dust whirl(s) or sand whirl(s), and no duststorm or sandstorm seen. **At sea**, visibility reduced by blowing spray.	
	08	Well developed dust whirl(s) or sand whirl(s) seen at or near the station during the preceding hour or at the time of observation, but no duststorm or sandstorm.	
	09	Duststorm or sandstorm within sight at the time of observation, or at the station during the preceding hour.	
Mist or shallow fog	10	Mist†, visibility 1,000 metres (0.5 n. miles) or more.	
	11	Shallow fog at the station, whether on land or sea, not deeper than about 2 metres on land or 10 metres (33 ft) at sea (visibility less than 1,000 metres — 0.5 n. miles).	in patches.
	12		more or less continuous.

* The expression "at the station" refers to a land station, a ship or an aircraft.
† See *Marine Observer's Guide*.

Table 12 — *continued*

ww = Present weather — *continued*

Special phenomena	13	Lightning visible, no thunder heard.	
	14	Precipitation within sight, not reaching the ground or surface of sea.	
	15	Precipitation within sight, reaching the ground or surface of sea, but distant (i.e. estimated to be more than 2.7 n. miles or 5 km) from station.	
	16	Precipitation within sight, reaching the ground or surface of sea, near to but not at the station.	
	17	Thunder audible during the 10 minutes preceding the time of observation, but no precipitation at the time of observation.	
	18	Squalls*	at or within sight of the station during the preceding hour or at the time of observation.
	19	Funnel cloud(s)†	

* A squall is defined as a sudden increase of wind speed by at least three stages of the Beaufort Scale, the speed rising to Force 6 or more and lasting for at least one minute.
† Tornado cloud or water spout.

ww = 20–29.   Precipitation, fog or thunderstorm at the station *during the preceding hour* **but not at the time of observation.**

20	Drizzle (not freezing) or snow grains.	
21	Rain (not freezing).	
22	Snow.	not falling as shower(s).
23	Rain and snow or ice pellets.	
24	Freezing drizzle or freezing rain.	
25	Shower(s) of rain.	
26	Shower(s) of snow, or of rain and snow.	
27	Shower(s) of hail, or of hail and rain.	
*28	Fog, visibility less than 1,000 metres (0.5 n. miles).	
†29	Thunderstorm (with or without precipitation or lightning).	

* When code figure 28 is used, the visibility must be greater than 1,000 metres a*t the time of observation.*
† Code figures 91–94 apply if there is precipitation *at the time of observation.*

Table 12 — *continued*

ww = Present weather — *continued*

ww = 30–39. **Duststorm, sandstorm, drifting or blowing snow.**

30	Slight or moderate duststorm or sandstorm	has decreased during the preceding hour.
31		no appreciable change during the preceding hour.
32		has begun or has increased during the preceding hour.
33	Severe duststorm or sandstorm	has decreased during the preceding hour.
34		no appreciable change during the preceding hour.
35		Has begun or has increased during the preceding hour.
36	Slight or moderate drifting snow	generally low (below eye level).
37	Heavy drifting snow	
38	Slight or moderate blowing snow	generally high (above eye level).
39	Heavy blowing snow	

ww = 40–49. **Fog* at the time of observation**

Visibility greater than 1,000 metres (0.5 n. miles)	40	Fog bank at a distance at the time of observation, but not at the station during the past hour†, the fog extending to a level above that of the observer.	
Visibility less than 1,000 metres (0.5 n. miles)	41	Fog in patches.	
	42	Fog, sky discernible	has become thinner during the preceding hour.
	43	Fog, sky not discernible	
	44	Fog, sky discernible	no appreciable change during the preceding hour.
	45	Fog, sky not discernible	
	46	Fog, sky discernible	has begun, or has become thicker during the preceding hour.
	47	Fog, sky not discernible	
	48	Fog, depositing rime	sky discernible.
	49		sky not discernible.

* Shallow fog at the station is covered by ww = 11, 12.
† Fog at the station during the past hour is covered by ww = 28.

Table 12 — *continued*

ww = Present weather — *continued*

ww = 50–99.   **PRECIPITATION* AT THE STATION AT THE TIME OF OBSERVATION**

ww = 50–59.   **Drizzle**

50	Drizzle, not freezing, intermittent	slight at time of observation.
51	Drizzle, not freezing, continuous	
52	Drizzle, not freezing, intermittent	moderate at time of observation.
53	Drizzle, not freezing, continuous	
54	Drizzle, not freezing, intermittent	heavy (dense) at time of observation.
55	Drizzle, not freezing, continuous	
56	Drizzle, freezing	slight.
57		moderate or heavy (dense).
58	Drizzle and rain	slight.
59		moderate or heavy.

ww = 60–69.   **Rain**

60	Rain, not freezing, intermittent	slight at time of observation.
61	Rain, not freezing, continuous	
62	Rain, not freezing, intermittent	moderate at time of observation.
63	Rain, not freezing, continuous	
64	Rain, not freezing, intermittent	heavy at time of observation.
65	Rain, not freezing, continuous	
66	Rain, freezing	slight.
67		moderate or heavy.
68	Rain or drizzle and snow	slight.
69		moderate or heavy.

* See also ww = 20–29, 30–35 for precipitation, etc. during preceding hour only. See also ww = 14–16 for precipitation away from the station.

Table 12 — *continued*

ww = Present weather — *continued*

ww = 70–79.   Solid precipitation, not in showers

70	Intermittent fall of snow flakes	slight at time of observation.
71	Continuous fall of snow flakes	
72	Intermittent fall of snow flakes	moderate at time of observation.
73	Continuous fall of snow flakes	
74	Intermittent fall of snow flakes	heavy at time of observation.
75	Continuous fall of snow flakes	
76	Ice prisms (with or without fog).	
77	Snow grains (with or without fog).	
78	Isolated star-like snow crystals (with or without fog).	
79	Ice pellets.	

ww = 80–90.   Showery precipitation

	80	Rain shower(s)	slight.
	81		moderate or heavy.
	82		violent.
No thunder at time of observation or in the preceding hour	83	Shower(s) of rain and snow mixed	slight.
	84		moderate or heavy.
	85	Snow shower(s)	slight.
	86		moderate or heavy.
	87	Shower(s) of snow pellets or ice pellets, with or without rain, or rain and snow mixed	slight.
	88		moderate or heavy.
	89	Shower(s) of hail, with or without rain, or rain and snow mixed, not associated with thunder	slight.
	90		moderate or heavy.

Table 12 — *continued*

ww = Present weather — *continued*

ww = 91–99.   Thunderstorm (either during the preceding hour or at the time of observation) with precipitation.

91	Slight rain at time of observation	Thunderstorm during the preceding hour, but not at time of observation.
92	Moderate or heavy rain at time of observation	
93	Slight snow, or rain and snow mixed, or hail at time of observation	
94	Moderate or heavy snow, or rain and snow mixed, or hail at time of observation	
95	Thunderstorm, slight or moderate, without hail, but with rain and/or snow at time of observation	Thunderstorm at the time of observation.
96	Thunderstorm, slight or moderate with hail at time of observation	
97	Thunderstorm, heavy, without hail, but with rain and/or snow at time of observation	
98	Thunderstorm combined with duststorm or sandstorm at time of observation	
99	Thunderstorm, heavy, with hail at time of observation	

Notes. (1) In general, the highest applicable figure should be selected. Only one exception should be mentioned: code figure ww = 17 has preference over all other code figures from 20 to 49 inclusive.

(2) The term "intermittent" applied to drizzle, rain or snow, used for various code figures between ww = 50 and 79, means that breaks have occurred during the preceding hour. Showers, however, are not included in this definition of intermittent precipitation but are to be reported by code figures between 80 and 99.

(3) "Continuous" precipitation means that it has lasted at least an hour without a break.

Table 13

W_1W_2 = Past weather

Code figure
- 0   Cloud cover ½ or less of the sky throughout the appropriate period.
- 1   Cloud cover ½ sky or less for part of the appropriate period and more than ½ sky for part of the period.
- 2   Cloud cover more than ½ of the sky throughout the appropriate period.
- 3   Duststorm, sand storm or blowing snow. ⎫ Visibility less
- 4   Fog or thick haze. ⎭ than 1,000 metres
- 5   Drizzle.
- 6   Rain.
- 7   Snow or rain and snow mixed.
- 8   Shower(s).
- 9   Thunder, with or without precipitation.

Notes. (1) This table refers to weather since the last synoptic hour (0000, 0600, 1200, 1800 UTC). For example, in a report made at 0600, W_1W_2 will cover the period 0000–0600. Ships reporting three hourly will report W_1W_2 for the three hour period.

(2) The code figures W_1W_2 are to be selected in such a way that W_1W_2 and ww together give as complete a description as possible of the weather in the time interval concerned; (e.g. if the type of weather undergoes a complete change during the time interval concerned, the code figures W_1W_2 should describe the weather prevailing before the type of weather indicated by ww began).

(3) Two code figures are applicable to the weather during the preceding 3 or 6 hours; the highest code figure is recorded under W_1 and the next highest is recorded for W_2. If the weather has been the same throughout the period, the code figures for W_1W_2 will be the same.

Table 14

C_L = Type of low cloud (Sc, St, Cu, Cb)

Code figure
- 0   No stratocumulus, stratus, cumulus or cumulonimbus.
- 1   Cumulus with little vertical extent and seemingly flattened, or ragged cumulus other than of bad weather*, or both.
- 2   Cumulus of moderate or strong vertical extent, generally with protuberances in the form of domes or towers. either accompanied or not by other cumulus or by stratocumulus, all having their bases at the same level.
- 3   Cumulonimbus the summits of which, at least partially, lack sharp outlines, but are neither clearly fibrous (cirriform) nor in the form of an anvil; cumulus, stratocumulus or stratus may also be present.
- 4   Stratocumulus formed by the spreading out of cumulus; cumulus may also be present.
- 5   Stratocumulus not resulting from the spreading out of cumulus.
- 6   Stratus in a more or less continuous sheet or layer, or in ragged shreds, or both, but no stratus fractus of bad weather.*
- 7   Stratus fractus of bad weather* or cumulus fractus of bad weather*, or both (pannus), usually below altostratus or nimbostratus.
- 8   Cumulus and stratocumulus other than that formed from the spreading out of cumulus; the base of the cumulus is at a different level from that of stratocumulus.

---

* "Bad weather" denotes the conditions which generally exist during precipitation and a short time before and after.

Table 14 — *continued*

C_L = Type of low cloud (Sc, St, Cu, Cb) — *continued*

9	Cumulonimbus, the upper part of which is clearly fibrous (cirriform), often in the form of an anvil; either accompanied or not by cumulonimbus without anvil or fibrous upper part, by cumulus, stratocumulus, stratus or pannus.
/	Stratocumulus, stratus, cumulus or cumulonimbus are invisible owing to fog, darkness or other surface phenomena.

Notes. (1) If there is fog but the sky is discernible through the fog, the cloud type, height and amount are reported as if no fog were present.

(2) In deciding which code figure to use when more than one cloud type is present, the order of priority, irrespective of quantity, is 9, 3, 4, 8, 2, otherwise whichever of the types 1, 5, 6 or 7 covers the largest area of sky.

Table 15

C_M = Type of medium cloud (Ac, As, Ns)

Code figure

0	No altocumulus, altostratus or nimbostratus.
1	Altostratus, the greater part of which is semi-transparent; through this part the sun or moon may be weakly visible, as through ground glass.
2	Altostratus, the greater part of which is sufficiently dense to hide the sun or moon, or nimbostratus.
3	Altocumulus, the greater part of which is semi-transparent; the various elements of the cloud change only slowly and are all at a single level.
4	Patches (often in the form of almonds or fishes) of altocumulus, the greater part of which is semi-transparent; the clouds occur at one or more levels and the elements are continually changing in appearance.
5	Semi-transparent altocumulus in bands, or altocumulus in one or more fairly continuous layers (semi-transparent or opaque), progressively invading the sky; these altocumulus clouds generally thicken as a whole.
6	Altocumulus resulting from the spreading out of cumulus (or cumulonimbus).
7	Altocumulus in two or more layers, usually opaque in places. and not progressively invading the sky; or opaque layer of altocumulus, not progressively invading the sky; or altocumulus together with altostratus or nimbostratus.
8	Altocumulus with sproutings in the form of small towers or battlements, or altocumulus having the appearance of cumuliform tufts.
9	Altocumulus of a chaotic sky, generally at several levels.
/	Altocumulus, altostratus or nimbostratus are invisible owing to fog, darkness or other surface phenomena, or because of the presence of a continuous layer of lower cloud.

Notes. (1) If there is fog but the sky is discernible through the fog, the cloud type, height and amount are reported as if no fog were present.

(2) In deciding which code figure to use when more than one cloud type is present, the order of priority should be: 9, 8, 7 whether altostratus or nimbostratus is present or not; then 6, 5, 4, 7, 3 if there is no altostratus and no nimbostratus; then 2, 1.

Table 16

C_H = Type of high cloud (Ci, Cc, Cs)

Code figure	
0	No cirrus, cirrocumulus or cirrostratus.
1	Cirrus in the form of filaments, strands or hooks, not progressively invading the sky.
2	Dense cirrus, in patches or entangled sheaves, which usually do not increase and sometimes seem to be the remains of the upper part of cumulonimbus; or cirrus with sproutings in the form of small turrets or battlements, or cirrus having the appearance of cumuliform tufts.
3	Dense cirrus, often in the form of an anvil, being the remains of the upper parts of cumulonimbus.
4	Cirrus in the form of hooks or of filaments, or both, progressively invading the sky; they generally become denser as a whole.
5	Cirrus (often in bands converging towards one point or two opposite points of the horizon) and cirrostratus, or cirrostratus alone; in either case, they are progressively invading the sky, and generally growing denser as a whole, but the continuous veil does not reach 45 degrees above the horizon.
6	Cirrus (often in bands converging towards one point or two opposite points of the horizon) and cirrostratus, or cirrostratus alone; in either case, they are progressively invading the sky, and generally growing denser as a whole; the continuous veil exceeds more than 45 degrees above the horizon, without the sky being totally covered.
7	Veil of cirrostratus covering the celestial dome.
8	Cirrostratus not progressively invading the sky and not completely covering the celestial dome.
9	Cirrocumulus alone, or cirrocumulus accompanied by cirrus or cirrostratus or both, but cirrocumulus is predominant.
/	Cirrus, cirrocumulus or cirrostratus are invisible owing to fog, darkness or other surface phenomena, or because of the presence of a continuous layer of lower cloud.

Notes. (1) If there is fog but the sky is discernible through the fog, the cloud type, height and amount are reported as if no fog were present.

(2) In deciding which code figure to use when more than one cloud type is present, the order of priority should be: 9, 7, 8, 6, 5, 4, 3, 2, 1.

Table 17

D_s = Ship's course (true) made good during the 3 hours preceding time of observation

Code figure	True Direction	Code figure	True Direction
0	Ship stopped	5	SW
1	NE	6	W
2	E	7	NW
3	SE	8	N
4	S	9	Unknown (course indirect)

Table 18

v_s = Ship's average speed made good during the 3 hours preceding time of observation

Code figure	Speed in knots	Code figure	Speed in knots
0	Ship stopped	5	21 to 25
1	1 to 5	6	26 to 30
2	6 to 10	7	31 to 35
3	11 to 15	8	36 to 40
4	16 to 20	9	Over 40

Table 19

I_s = Type of ice accretion

Code figure	Description
1	Icing from sea spray.
2	Icing from fog.
3	Icing from spray and fog.
4	Icing from rain.
5	Icing from spray and rain.

Table 20

R_s = Rate of ice accretion

Code figure	
0	Ice not building up.
1	Ice building up slowly.
2	Ice building up rapidly.
3	Ice melting or breaking up slowly.
4	Ice melting or breaking up rapidly.

Table 21

c_i = Concentration or arrangement of sea ice.

Code figure		
0	No ice.	
1	Ship in open lead more than 1 n. mile wide or ship in fast ice with boundary beyond limit of visibility.	
2–5	**Ice concentration uniform.**	
2	Open water or very open pack ice, < 3/8 concentration.	Ship in ice or within 0.5 n. mile of ice
3	Open pack ice 3/8 to < 6/8 concentration.	
4	Close pack ice 6/8 to < 7/8 concentration.	
5	Very close pack ice 7/8 < 8/8 concentration.	

Table 21 — *continued*

6–9 Ice concentration not uniform
6   Strips and patches of pack ice with open water between.
7   Strips and patches of close or very close pack ice with areas of lesser concentration between.
8   Fast ice with open water, very open or open pack ice to seaward of the ice boundary.
9   Fast ice with close or very close pack ice to seaward of the ice boundary.

⎱ Ship in ice or within 0.5 n.mile of ice

/   Unable to report, because of darkness, poor visibility or ship is more than 0.5 n. mile away from ice edge.

Table 22

S_i = Stage of development

Code figure
0   New ice only (frazil ice, grease ice. slush, shuga).
1   Nilas or ice rind, < 10cm thick.
2   Young ice (grey ice, grey-white ice) 10–30 cm thick.
3   Predominantly new and/or young ice with some first-year ice.
4   Predominantly thin first-year ice with some new and/or young ice.
5   All thin first-year ice (30–70 cm thick).
6   Predominantly medium first-year ice (70–120 cm thick) and thick first-year ice (> 120 cm thick) and some thinner (younger) first-year ice.
7   All medium and thick first-year ice.
8   Predominantly medium and thick first-year ice with some old ice (usually more than 2 metres thick).
9   Predominantly old ice.
/   Unable to report, because of darkness, poor visibility, only ice of land origin visible or ship is more than 0.5 nautical miles away from ice edge.

Table 23

b_i = Ice of land origin

Code figure
0   No ice of land origin.
1   1–5 icebergs, no growlers or bergy bits.
2   6–10 icebergs. no growlers or bergy bits.
3   11–20 icebergs, no growlers or bergy bits.
4   Up to and including 10 growlers and bergy bits — no icebergs.
5   More than 10 growlers and bergy bits — no icebergs.
6   1–5 icebergs with growlers and bergy bits.
7   6–10 icebergs with growlers and bergy bits.
8   11–20 icebergs with growlers and bergy bits.
9   More than 20 icebergs with growlers and bergy bits — a major hazard to navigation.
/   Unable to report — because of darkness, poor visibility or only sea ice is visible.

Note. If only ice of land origin is visible the group should be reported as $0/b_i/0$ e.g. 0/2/0 would mean 6–10 icebergs in sight.

Table 24
D_i = Bearing of principal ice edge

Code figure
- 0   Ship in shore or flaw lead.
- 1   Principal ice edge towards North-East.
- 2   Principal ice edge towards East.
- 3   Principal ice edge towards South-East.
- 4   Principal ice edge towards South.
- 5   Principal ice edge towards South-West.
- 6   Principal ice edge towards West.
- 7   Principal ice edge towards North-West.
- 8   Principal ice edge towards North.
- 9   Not determined (ship in ice).
- /   Unable to report — because of darkness, poor visibility or only ice of land origin is visible.

Note. If the ship is more than 0.5 n.mile from an ice edge the group should be reported as ///D_i0 e.g. ice edge to the North would be ///80.

Table 25
z_i = Ice situation and trend over preceding 3 hours

Code figure
- 0   Ship in open water with floating ice in sight.
- 1   Ship in easily penetrable ice; conditions improving.
- 2   Ship in easily penetrable ice, conditions not changing.
- 3   Ship in easily penetrable ice, conditions worsening.
- 4   Ship in ice difficult to penetrate; conditions improving.
- 5   Ship in ice difficult to penetrate; conditions not changing.
- **6–9   Ice difficult to penetrate, conditions worsening.**
- 6   Ice forming and floes freezing together.
- 7   Ice under slight pressure.
- 8   Ice under moderate or severe pressure.
- 9   Ship beset.
- /   Unable to report—because of darkness or poor visibility.

(Codes 1–9: **Ship in ice**)

Table 26
s_s = Indicator for sign and type of measurement of sea surface temperature

Code figure	Sign	Type of measurement	Code figure	Sign	Type of measurement
0	positive or 0	intake	5	negative	hull contact sensor
1	negative	intake	6	positive or 0	other
2	positive or 0	bucket	7	negative	other
3	negative	bucket			
4	positive or 0	hull contact sensor (i.e. Met. Office supply)			

Table 27
s_w = Indicator for the sign and type of wet bulb temperature reported

Code figure	Sign	Type of measurement	Code figure	Sign	Type of measurement
0	positive or 0	measured	5	positive or 0	computed
1	negative	measured	6	negative	computed
2	iced bulb	measured	7	iced bulb	computed

Table for finding the dew-point (°C)

(For use with Marine Screen)

DRY BULB	\multicolumn{19}{c	}{Depression of wet bulb}	DRY BULB
	0.0	0.2	0.4	0.6	0.8	1.0	1.2	1.4	1.6	1.8	2.0	2.5	3.0	3.5	4.0	4.5	5.0	5.5	6.0	6.5	7.0	7.5	8.0	8.5	9.0	
40.0	40.0	39.8	39.5	39.3	39.0	38.8	38.5	38.3	38.0	37.8	37.5	36.9	36.3	35.6	34.9	34.3	33.6	32.9	32.3	31.6	30.9	30.1	29.4	28.7	27.9	40.0
39.5	39.5	39.3	39.0	38.8	38.5	38.3	38.0	37.8	37.5	37.3	37.0	36.4	35.8	35.1	34.4	33.8	33.1	32.4	31.8	31.0	30.3	29.6	28.9	28.1	27.3	39.5
39.0	39.0	38.8	38.5	38.3	38.0	37.8	37.5	37.3	37.0	36.8	36.5	35.8	35.3	34.6	33.9	33.3	32.6	31.9	31.2	30.5	29.8	29.0	28.3	27.5	26.8	39.0
38.5	38.5	38.3	38.0	37.8	37.5	37.3	37.0	36.8	36.5	36.3	36.0	35.3	34.7	34.0	33.4	32.7	32.0	31.3	30.6	29.9	29.2	28.5	27.8	26.9	26.2	38.5
38.0	38.0	37.8	37.5	37.3	37.0	36.8	36.5	36.3	36.0	35.8	35.5	34.8	34.2	33.5	32.8	32.2	31.5	30.8	30.1	29.4	28.6	27.9	27.1	26.4	25.6	38.0
37.5	37.5	37.3	37.0	36.8	36.5	36.3	36.0	35.8	35.5	35.2	34.9	34.3	33.6	32.9	32.3	31.6	30.9	30.3	29.5	28.8	28.1	27.3	26.6	25.8	25.0	37.5
37.0	37.0	36.8	36.5	36.3	36.0	35.8	35.5	35.2	34.9	34.7	34.4	33.8	33.1	32.5	31.8	31.1	30.4	29.8	29.0	28.3	27.5	26.8	26.0	25.2	24.4	37.0
36.5	36.5	36.3	36.0	35.8	35.5	35.2	34.9	34.7	34.4	34.2	33.9	33.3	32.6	31.9	31.3	30.6	29.9	29.2	28.4	27.7	26.9	26.2	25.4	24.6	23.8	36.5
36.0	36.0	35.8	35.4	35.2	34.9	34.7	34.4	34.2	33.9	33.7	33.4	32.8	32.1	31.4	30.8	30.0	29.3	28.6	27.9	27.1	26.4	25.6	24.8	24.0	23.2	36.0
35.5	35.5	35.2	34.9	34.7	34.4	34.2	33.9	33.7	33.4	33.2	32.9	32.3	31.6	30.9	30.2	29.5	28.8	28.1	27.3	26.6	25.8	25.0	24.3	23.4	22.6	35.5
35.0	35.0	34.7	34.4	34.2	33.9	33.7	33.4	33.2	32.9	32.6	32.4	31.8	31.0	30.4	29.7	29.0	28.3	27.5	26.8	26.0	25.3	24.4	23.6	22.8	21.9	35.0
34.5	34.5	34.2	33.9	33.7	33.4	33.2	32.9	32.7	32.4	32.1	31.9	31.2	30.5	29.8	29.1	28.4	27.7	27.0	26.3	25.4	24.7	23.9	23.0	22.2	21.3	34.5
34.0	34.0	33.7	33.4	33.2	32.9	32.7	32.4	32.1	31.9	31.6	31.3	30.7	30.0	29.3	28.6	27.9	27.1	26.4	25.6	24.9	24.1	23.3	22.4	21.6	20.7	34.0
33.5	33.5	33.2	32.9	32.7	32.4	32.2	31.9	31.6	31.4	31.1	30.8	30.2	29.5	28.8	28.1	27.3	26.6	25.8	25.1	24.3	23.5	22.7	21.8	21.0	20.1	33.5
33.0	33.0	32.7	32.4	32.2	31.9	31.7	31.4	31.1	30.9	30.6	30.3	29.6	28.9	28.3	27.5	26.8	26.0	25.3	24.5	23.8	22.9	22.1	21.3	20.3	19.4	33.0
32.5	32.5	32.2	31.9	31.7	31.4	31.1	30.9	30.6	30.3	30.1	29.8	29.1	28.4	27.8	27.0	26.3	25.5	24.8	23.9	23.1	22.3	21.5	20.6	19.8	18.8	32.5
32.0	32.0	31.7	31.4	31.2	30.9	30.6	30.4	30.1	29.8	29.6	29.3	28.6	27.9	27.2	26.4	25.7	24.9	24.2	23.4	22.6	21.8	20.9	20.0	19.1	18.2	32.0
31.5	31.5	31.2	30.9	30.7	30.4	30.1	29.9	29.6	29.3	29.0	28.8	28.1	27.4	26.6	25.9	25.2	24.4	23.6	22.8	22.0	21.1	20.3	19.4	18.5	17.5	31.5
31.0	31.0	30.7	30.4	30.2	29.9	29.6	29.3	29.1	28.8	28.5	28.3	27.5	26.8	26.1	25.4	24.6	23.8	23.0	22.3	21.4	20.5	19.7	18.8	17.8	16.8	31.0
30.5	30.5	30.2	29.9	29.7	29.4	29.1	28.8	28.6	28.3	28.0	27.8	27.0	26.3	25.6	24.8	24.1	23.3	22.5	21.6	20.8	19.9	19.0	18.1	17.2	16.2	30.5
30.0	30.0	29.7	29.4	29.1	28.9	28.6	28.3	28.1	27.8	27.5	27.3	26.5	25.8	25.0	24.3	23.5	22.8	21.9	21.1	20.3	19.3	18.4	17.5	16.5	15.6	30.0
29.5	29.5	29.2	28.9	28.6	28.4	28.1	27.8	27.5	27.3	27.0	26.7	26.0	25.3	24.5	23.8	22.9	22.1	21.3	20.5	19.6	18.8	17.8	16.8	15.9	14.9	29.5
29.0	29.0	28.7	28.4	28.1	27.9	27.6	27.3	27.0	26.8	26.5	26.2	25.4	24.8	23.9	23.2	22.4	21.6	20.8	19.9	19.0	18.1	17.2	16.3	15.2	14.2	29.0
28.5	28.5	28.2	27.9	27.6	27.4	27.1	26.8	26.5	26.3	25.9	25.7	24.9	24.2	23.4	22.6	21.8	21.0	20.2	19.3	18.4	17.5	16.5	15.6	14.6	13.5	28.5
28.0	28.0	27.7	27.4	27.1	26.8	26.6	26.3	26.0	25.8	25.4	25.1	25.1	23.6	22.9	22.1	21.3	20.4	19.6	18.8	17.8	16.9	15.9	14.9	13.9	12.8	28.0
27.5	27.5	27.2	26.9	26.6	26.3	26.1	25.8	25.5	25.2	24.9	24.6	23.9	23.1	22.3	21.5	20.8	19.9	19.0	18.1	17.2	16.3	15.3	14.3	13.2	12.1	27.5
27.0	27.0	26.7	26.4	26.1	25.8	25.5	25.3	25.0	24.7	24.4	24.1	23.3	22.6	21.8	21.0	20.1	19.3	18.4	17.5	16.6	15.7	14.7	13.6	12.5	11.4	27.0
26.5	26.5	26.2	25.9	25.6	25.3	25.0	24.8	24.5	24.2	23.9	23.6	22.8	22.0	21.3	20.4	19.6	18.8	17.8	16.9	16.0	15.0	14.0	12.9	11.8	10.6	26.5
26.0	26.0	25.7	25.4	25.1	24.8	24.5	24.3	23.9	23.6	23.3	23.0	22.2	21.5	20.7	19.9	19.0	18.1	17.3	16.3	15.4	14.4	13.3	12.2	11.1	9.9	26.0
25.5	25.5	25.2	24.9	24.6	24.3	24.0	23.8	23.4	23.1	22.8	22.5	21.8	20.9	20.1	19.3	18.4	17.5	16.6	15.7	14.7	13.7	12.7	11.5	10.4	9.1	25.5
25.0	25.0	24.7	24.4	24.1	23.8	23.5	23.3	22.9	22.6	22.3	22.0	21.3	20.4	19.6	18.8	17.9	17.0	16.0	15.1	14.1	13.1	12.0	10.8	9.6	8.4	25.0

Dew-point tables — *continued*

DRY BULB	\multicolumn{19}{c}{Depression of wet bulb}																			DRY BULB
	0.0	0.2	0.4	0.6	0.8	1.0	1.2	1.4	1.6	1.8	2.0	2.5	3.0	3.5	4.0	4.5	5.0	5.5	6.0	6.5	7.0	7.5	8.0	8.5	9.0	
24.5	24.5	24.2	23.9	23.6	23.3	23.0	22.7	22.4	22.1	21.8	21.5	20.7	19.9	19.0	18.2	17.3	16.4	15.5	14.5	13.5	12.4	11.3	10.1	8.9	7.6	24.5
24.0	24.0	23.7	23.4	23.1	22.8	22.5	22.2	21.9	21.6	21.3	20.9	20.1	19.3	18.5	17.6	16.8	15.8	14.9	13.9	12.8	11.7	10.6	9.4	8.1	6.8	24.0
23.5	23.5	23.2	22.9	22.6	22.3	22.0	21.7	21.3	21.0	20.8	20.4	19.6	18.8	17.9	17.0	16.1	15.2	14.2	13.2	12.2	11.1	9.9	8.7	7.3	5.9	23.5
23.0	23.0	22.7	22.4	22.1	21.8	21.4	21.1	20.8	20.5	20.2	19.9	19.1	18.3	17.3	16.5	15.6	14.6	13.6	12.6	11.5	10.4	9.2	7.9	6.6	5.1	23.0
22.5	22.5	22.2	21.9	21.5	21.3	20.9	20.6	20.3	20.0	19.7	19.3	18.5	17.7	16.8	15.9	15.0	14.0	13.0	11.9	10.8	9.7	8.4	7.1	5.7	4.3	22.5
22.0	22.0	21.6	21.3	21.0	20.8	20.4	20.1	19.8	19.5	19.1	18.8	18.0	17.1	16.3	15.4	14.4	13.4	12.4	11.3	10.2	9.0	7.7	6.4	4.9	3.4	22.0
21.5	21.5	21.1	20.8	20.5	20.3	19.9	19.6	19.3	18.9	18.6	18.3	17.4	16.6	15.7	14.8	13.8	12.8	11.7	10.6	9.5	8.3	7.0	5.6	4.1	2.5	21.5
21.0	21.0	20.6	20.3	20.0	19.8	19.4	19.1	18.8	18.4	18.1	17.8	16.9	16.0	15.1	14.2	13.2	12.2	11.1	10.0	8.8	7.5	6.2	4.8	3.2	1.5	21.0
20.5	20.5	20.1	19.8	19.5	19.2	18.9	18.5	18.3	17.9	17.6	17.3	16.3	15.5	14.6	13.6	12.6	11.5	10.5	9.3	8.1	6.8	5.4	3.9	2.3	0.6	20.5
20.0	20.0	19.6	19.3	19.0	18.7	18.4	18.0	17.7	17.4	17.0	16.7	15.8	14.9	14.0	13.0	12.0	10.9	9.8	8.6	7.4	6.0	4.6	3.1	1.4	−0.4	20.0
19.5	19.5	19.1	18.8	18.5	18.2	17.8	17.5	17.2	16.8	16.5	16.1	15.3	14.4	13.4	12.4	11.4	10.3	9.1	7.9	6.6	5.3	3.8	2.2	0.5	−1.4	19.5
19.0	19.0	18.6	18.3	18.0	17.7	17.3	17.0	16.6	16.3	16.0	15.6	14.7	13.8	12.8	11.8	10.8	9.6	8.5	7.2	5.9	4.5	3.0	1.3	−0.5	−2.5	19.0
18.5	18.5	18.1	17.8	17.5	17.1	16.8	16.5	16.1	15.8	15.5	15.1	14.2	13.2	12.2	11.2	10.1	9.0	7.8	6.5	5.2	3.7	2.1	0.4	−1.5	−3.6	18.5
18.0	18.0	17.6	17.3	17.0	16.6	16.3	16.0	15.6	15.3	14.9	14.6	13.6	12.7	11.7	10.6	9.5	8.3	7.1	5.8	4.4	2.9	1.3	−0.5	−2.5	−4.8	18.0
17.5	17.5	17.1	16.8	16.5	16.1	15.8	15.5	15.1	14.8	14.4	14.0	13.1	12.1	11.1	10.0	8.9	7.7	6.4	5.1	3.6	2.1	0.4	−1.5	−3.6	−6.0	17.5
17.0	17.0	16.6	16.3	16.0	15.6	15.3	14.9	14.6	14.2	13.9	13.5	12.5	11.5	10.5	9.4	8.2	7.0	5.7	4.3	2.9	1.2	−0.5	−2.5	−4.7	−7.3	17.0
16.5	16.5	16.1	15.8	15.5	15.1	14.8	14.4	14.1	13.7	13.3	12.9	12.0	10.9	9.9	8.8	7.6	6.3	5.0	3.6	2.0	0.4	−1.5	−3.6	−5.9	−8.7	16.5
16.0	16.0	15.7	15.3	15.0	14.6	14.3	13.9	13.5	13.2	12.8	12.4	11.4	10.4	9.3	8.1	6.9	5.7	4.3	2.8	1.2	−0.5	−2.5	−4.6	−7.2	−10.2	16.0
15.5	15.5	15.2	14.8	14.5	14.1	13.7	13.4	13.0	12.6	12.2	11.8	10.8	9.8	8.7	7.5	6.3	5.0	3.6	2.0	0.4	−1.4	−3.5	−5.8	−8.5	−11.8	15.5
15.0	15.0	14.7	14.3	13.9	13.6	13.2	12.8	12.5	12.1	11.7	11.3	10.3	9.2	8.1	6.9	5.6	4.3	2.8	1.2	−0.5	−2.4	−4.5	−7.0	−9.9	−13.5	15.0
14.5	14.5	14.1	13.8	13.4	13.1	12.7	12.3	11.9	11.5	11.1	10.7	9.7	8.6	7.4	6.2	4.9	3.5	2.0	0.4	−1.4	−3.4	−5.6	−8.3	−11.4	−15.5	14.5
14.0	14.0	13.6	13.3	12.9	12.5	12.2	11.8	11.4	11.0	10.6	10.2	9.1	8.0	6.8	5.6	4.2	2.8	1.3	−0.4	−2.3	−4.4	−6.8	−9.6	−13.1	−17.7	14.0
13.5	13.5	13.1	12.8	12.4	12.0	11.6	11.2	10.9	10.4	10.0	9.6	8.5	7.4	6.2	4.9	3.6	2.1	0.5	−1.3	−3.2	−5.4	−8.0	−11.0	−14.9	−20.2	13.5
13.0	13.0	12.6	12.3	11.9	11.5	11.1	10.7	10.3	9.9	9.5	9.1	8.0	6.8	5.6	4.3	2.9	1.3	−0.3	−2.1	−4.2	−6.5	−9.3	−12.6	−16.9	−23.3	13.0
12.5	12.5	12.1	11.8	11.4	11.0	10.6	10.2	9.8	9.4	8.9	8.5	7.4	6.2	4.9	3.6	2.1	0.6	−1.1	−3.0	−5.2	−7.7	−10.6	−14.3	−19.3	−27.3	12.5
12.0	12.0	11.6	11.2	10.9	10.5	10.1	9.6	9.2	8.8	8.4	7.9	6.8	5.6	4.3	2.9	1.4	−0.2	−2.0	−4.0	−6.2	−8.9	−12.1	−16.2	−22.0	−33.0	12.0
11.5	11.5	11.1	10.7	10.3	9.9	9.5	9.1	8.7	8.3	7.8	7.4	6.2	5.0	3.6	2.2	0.7	−1.0	−2.9	−4.9	−7.3	−10.2	−13.6	−18.3	−25.3	−44.4	11.5
11.0	11.0	10.6	10.2	9.8	9.4	9.0	8.6	8.1	7.7	7.3	6.8	5.6	4.3	3.0	1.5	−0.1	−1.8	−3.7	−5.9	−8.5	−11.5	−15.4	−20.7	−29.8		11.0
10.5	10.5	10.1	9.7	9.3	8.9	8.5	8.0	7.6	7.2	6.7	6.2	5.0	3.7	2.3	0.8	−0.8	−2.6	−4.7	−7.0	−9.7	−13.0	−17.3	−23.6	−36.8		10.5
10.0	10.0	9.6	9.2	8.8	8.4	7.9	7.5	7.1	6.6	6.1	5.7	4.4	3.1	1.6	0.1	−1.6	−3.5	−5.6	−8.1	−11.0	−14.6	−19.4	−27.2			10.0
9.5	9.5	9.1	8.7	8.3	7.8	7.4	7.0	6.5	6.0	5.6	5.1	3.8	2.4	1.0	−0.6	−2.4	−4.4	−6.6	−9.2	−12.3	−16.3	−21.9	−32.2			9.5
9.0	9.0	8.6	8.2	7.7	7.3	6.9	6.4	6.0	5.5	5.0	4.5	3.2	1.8	0.3	−1.4	−3.2	−5.3	−7.6	−10.4	−13.8	−18.2	−24.9	−40.8			9.0
8.5	8.5	8.1	7.7	7.2	6.8	6.3	5.9	5.4	4.9	4.4	3.9	2.6	1.1	−0.4	−2.1	−4.0	−6.2	−8.7	−11.6	−15.3	−20.4	−28.7				8.5
8.0	8.0	7.6	7.1	6.7	6.3	5.8	5.3	4.8	4.3	3.8	3.3	2.0	0.5	−1.1	−2.9	−4.9	−7.1	−9.8	−13.0	−17.0	−22.9	−34.2				8.0

Dew-point tables — continued

DRY BULB	Depression of wet bulb																			DRY BULB
	0.0	0.2	0.4	0.6	0.8	1.0	1.2	1.4	1.6	1.8	2.0	2.5	3.0	3.5	4.0	4.5	5.0	5.5	6.0	6.5	7.0	7.5	8.0	8.5	9.0	
7.5	7.5	7.1	6.6	6.2	5.7	5.3	4.8	4.3	3.8	3.3	2.7	1.3	−0.2	−1.9	−3.7	−5.8	−8.1	−10.9	−14.4	−19.0	−26.0	−44.7				7.5
7.0	7.0	6.6	6.1	5.7	5.2	4.7	4.2	3.7	3.2	2.7	2.1	0.7	−0.9	−2.6	−4.5	−6.7	−9.2	−1.2	−15.9	−21.1	−29.9	−31.1				7.0
6.5	6.5	6.1	5.6	5.1	4.7	4.2	3.7	3.2	2.6	2.1	1.5	0.1	−1.6	−3.3	−5.3	−7.6	−10.2	−13.4	−17.6	−23.6	−24.8	−38.4				6.5
6.0	6.0	5.5	5.1	4.6	4.1	3.6	3.1	2.6	2.1	1.5	0.9	−0.6	−2.3	−4.1	−6.2	−8.5	−11.4	−14.8	−19.5	−20.7	−28.6					6.0
5.5	5.5	5.0	4.6	4.1	3.6	3.1	2.6	2.0	1.5	0.9	0.3	−1.2	−3.0	−4.9	−7.0	−9.5	−12.5	−16.3	−17.6	−23.3	−33.8					5.5
5.0	5.0	4.5	4.1	3.6	3.1	2.5	2.0	1.5	0.9	0.3	−0.3	−1.9	−3.7	−5.7	−7.9	−10.6	−13.8	−15.1	−19.6	−26.4	−43.4					5.0
4.5	4.5	4.0	3.5	3.0	2.5	2.0	1.4	0.9	0.3	−0.3	−0.9	−2.6	−4.4	−6.5	−8.8	−11.6	−13.0	−16.7	−21.8	−30.3						4.5
4.0	4.0	3.5	3.0	2.5	2.0	1.4	0.9	0.3	−0.3	−0.9	−1.5	−3.2	−5.1	−7.3	−9.8	−11.1	−14.3	−18.4	−24.4	−36.1						4.0
3.5	3.5	3.0	2.5	2.0	1.4	0.9	0.3	−0.3	−0.9	−1.5	−2.2	−3.9	−5.9	−8.1	−9.5	−12.3	−15.8	−20.4	−27.5	−48.4						3.5
3.0	3.0	2.5	2.0	1.4	0.9	0.3	−0.3	−0.9	−1.5	−2.1	−2.8	−4.6	−6.7	−8.0	−10.5	−13.5	−17.3	−22.6	−31.6							3.0
2.5	2.5	2.0	1.5	0.9	0.4	−0.2	−0.8	−1.4	−2.1	−2.8	−3.5	−5.3	−6.7	−9.0	−11.6	−14.9	−19.0	−25.1	−37.7							2.5
2.0	2.0	1.5	1.0	0.4	−0.2	−0.8	−1.4	−2.0	−2.7	−3.4	−4.1	−5.5	−7.6	−10.0	−12.8	−16.3	−20.9	−28.2	−51.9							2.0
1.5	1.5	1.0	0.4	−0.2	−0.7	−1.3	−2.0	−2.6	−3.0	−3.7	−4.4	−6.3	−8.4	−11.0	−14.0	−17.8	−23.1	−32.2								1.5
1.0	1.0	0.5	−0.1	−0.7	−1.3	−1.9	−2.3	−3.0	−3.7	−4.4	−5.1	−7.1	−9.3	−12.0	−15.2	−19.4	−25.5	−38.1								1.0
0.5	0.5	−0.1	−0.6	−1.1	−1.7	−2.4	−3.0	−3.7	−4.4	−5.1	−5.8	−7.9	−10.3	−13.1	−16.6	−21.2	−28.4	−51.4								0.5
0.0	0.0	−0.6	−1.1	−1.7	−2.3	−3.0	−3.6	−4.3	−5.1	−5.8	−6.6	−8.7	−11.2	−14.2	−18.0	−23.2	−32.2									0.0
−0.5	−0.6	−1.1	−1.7	−2.3	−3.0	−3.6	−4.3	−5.0	−5.8	−6.5	−7.4	−9.6	−12.2	−15.4	−19.5	−25.5	−37.4									
−1.0	−.1	−1.7	−2.3	−2.9	−3.6	−4.3	−5.0	−5.7	−6.5	−7.3	−8.1	−10.5	−13.2	−16.6	−21.2	−28.1	−47.5									
−1.5	−.7	−2.3	−2.9	−3.6	−4.2	−4.9	−5.6	−6.4	−7.2	−8.0	−8.9	−11.4	−14.3	−17.9	−23.0	−31.4										
−2.0	−2.3	−2.9	−3.5	−4.2	−4.9	−5.6	−6.3	−7.1	−7.9	−8.8	−9.7	−12.3	−15.4	−19.3	−25.0	−35.8										
−2.5	−2.8	−3.5	−4.1	−4.8	−5.5	−6.2	−7.0	−7.8	−8.7	−9.6	−10.5	−13.2	−16.5	−20.9	−27.4	−42.9										
−3.0	−3.4	−4.0	−4.7	−5.4	−6.1	−6.9	−7.7	−8.5	−9.4	−10.3	−11.3	−14.2	−17.7	−22.5	−30.2											
−3.5	−4.0	−4.6	−5.3	−6.0	−6.8	−7.5	−8.4	−9.2	−10.2	−11.1	−12.2	−15.2	−19.0	−24.3	−33.7											
−4.0	−4.5	−5.2	−5.9	−6.6	−7.4	−8.2	−9.1	−10.0	−10.9	−11.9	−13.0	−16.2	−20.3	−26.3	−38.7											
−4.5	−5.1	−5.8	−6.5	−7.3	−8.0	−8.9	−9.8	−10.7	−11.7	−12.8	−13.9	−17.3	−21.8	−28.7	−47.5											
−5.0	−5.6	−6.4	−7.1	−7.9	−8.7	−9.6	−10.5	−11.4	−12.5	−13.6	−14.8	−18.4	−23.3	−31.4												
−5.5	−6.2	−6.9	−7.7	−8.5	−9.3	−10.2	−11.2	−12.2	−13.3	−14.5	−15.6	−19.6	−25.0	−34.9												
−6.0	−6.8	−7.5	−8.3	−9.1	−10.0	−10.9	−11.9	−13.0	−14.1	−15.4	−16.7	−20.9	−26.9	−39.7												
−6.5	−7.3	−8.1	−8.9	−9.8	−10.7	−11.6	−12.7	−13.8	−15.0	−16.3	−17.7	−22.2	−29.1	−48.4												
−7.0	−7.9	−8.7	−9.5	−10.4	−11.3	−12.3	−13.4	−14.6	−15.8	−17.2	−18.7	−23.6	−31.6													
−7.5	−8.4	−9.3	−10.1	−11.0	−12.0	−13.0	−14.2	−15.4	−16.7	−18.2	−19.8	−25.1	−34.7													
−8.0	−9.0	−9.8	−10.7	−11.7	−12.7	−13.8	−14.9	−16.2	−17.6	−19.2	−20.9	−26.8	−38.7													
−8.5	−9.6	−10.4	−11.3	−12.3	−13.4	−14.5	−15.7	−17.0	−18.5	−20.2	−22.1	−28.7	−45.0													
−9.0	−10.1	−11.0	−11.9	−13.0	−14.0	−15.2	−16.5	−17.9	−19.5	−21.3	−23.3	−30.8														

Dew-point tables — *continued*

| DRY BULB | \multicolumn{16}{c|}{Depression of wet bulb} | DRY BULB |

DRY BULB	0.0	0.2	0.4	0.6	0.8	1.0	1.2	1.4	1.6	1.8	2.0	2.5	3.0	3.5	4.0	4.5	5.0	5.5	6.0	6.5	7.0	7.5	8.0	8.5	9.0
−9.5	−10.7	−11.6	−12.6	−13.6	−14.7	−16.0	−17.3	−18.8	−20.5	−22.4	−24.7	−33.3													
−10.0	−11.2	−12.2	−13.2	−14.3	−15.4	−16.7	−18.1	−19.7	−21.5	−23.6	−26.1	−36.3													
−10.5	−11.8	−12.8	−13.8	−14.9	−16.2	−17.5	−19.0	−20.7	−22.6	−24.9	−27.6	−40.3													
−11.0	−12.3	−13.3	−14.4	−15.6	−16.9	−18.3	−19.9	−21.7	−23.7	−26.2	−29.3	−46.4													
−11.5	−12.9	−13.9	−15.1	−16.3	−17.6	−19.1	−20.8	−22.7	−24.9	−27.6	−31.2														
−12.0	−13.4	−14.5	−15.7	−17.0	−18.4	−19.9	−21.7	−23.7	−26.2	−29.2	−33.3														
−12.5	−14.0	−15.1	−16.3	−17.6	−19.1	−20.7	−22.6	−24.8	−27.5	−30.9	−35.8														
−13.0	−14.6	−15.7	−17.0	−18.3	−19.9	−21.6	−23.6	−26.0	−29.0	−32.9	−38.9														
−13.5	−15.1	−16.3	−17.6	−19.0	−20.6	−22.5	−24.6	−27.2	−30.5	−35.1	−43.1														
−14.0	−15.7	−16.9	−18.2	−19.7	−21.4	−23.4	−25.7	−28.5	−32.3	−37.8	−49.8														
−14.5	−16.2	−17.5	−18.9	−20.5	−22.3	−24.3	−26.8	−29.9	−34.2	−41.1															
−15.0	−16.8	−18.1	−19.5	−21.2	−23.1	−25.3	−28.0	−31.5	−36.5	−45.9															
−15.5	−17.3	−18.7	−20.2	−21.9	−23.9	−26.3	−29.2	−33.2	−39.2																
−16.0	−17.9	−19.3	−20.9	−22.7	−24.8	−27.4	−30.6	−35.0	−42.6																
−16.5	−18.4	−19.9	−21.5	−23.5	−25.7	−28.5	−32.0	−37.2	−47.5																
−17.0	−19.0	−20.5	−22.2	−24.2	−26.6	−29.6	−33.6	−39.8																	
−17.5	−19.5	−21.1	−22.9	−25.0	−27.6	−30.8	−35.3	−43.0																	
−18.0	−20.1	−21.7	−23.6	−25.9	−28.6	−32.1	−37.3	−47.5																	
−18.5	−20.6	−22.3	−24.3	−26.7	−29.6	−33.5	−39.6																		
−19.0	−21.1	−22.9	−25.0	−27.5	−30.7	−35.1	−42.3																		
−19.5	−21.7	−23.6	−25.8	−28.4	−31.9	−36.8	−45.9																		
−20.0	−22.2	−24.2	−26.5	−29.3	−33.1	−38.7	−51.3																		

In the tables, lines are ruled to draw attention to the fact that above the line evaporation is going on from a water surface, while below the line it is going on from an ice surface (wet-bulb temperature below 0 °C). Owing to this, interpolation must not be made between figures on different sides of the line.

For dry bulb temperatures below 0 °C, it will be noticed that, when the depression of the wet bulb is zero, i.e. when the temperature of the wet bulb is equal to that of the dry bulb, the dew-point may be below the dry bulb, so that the relative humidity is less than 100 per cent. These apparent anomalies are a consequence of the method of computing dew-points and relative humidities now adopted by the Meteorological Office, in which, for temperatures below 0 °C, the dew-point is the temperature at which the actual vapour pressure is equal to the saturation vapour pressure over supercooled water (not over ice) and the relative humidity is the ratio, expressed as a percentage, of the actual vapour pressure to the saturation vapour pressure over supercooled water at the temperature of the dry bulb.

SECTION 4
THE ATLANTIC WEATHER BULLETIN*

This bulletin is broadcast for the purpose of giving mariners all available information about the weather existing in the eastern North Atlantic in a most concise manner. Similar bulletins are issued for other ocean areas by the relevant National Meteorological Services. The bulletin is broadcast for the benefit of those desiring only part of the available information.

The bulletin consists of three parts, 1, 2 and 3, broadcast from Portishead Radio at 0930 and 2130 UTC.

Contents of the Atlantic Weather Bulletin

Part 1. Storm Warnings (wind force 10 or more) in plain language. Warnings contain information about the nature, depth, extent, movement and expected development of the disturbance causing the storm. When there are no storms in the forecast area, that fact is indicated by the words 'No storm warnings'.

Part 2. General Synoptic Development.

Part 3. Forecasts in plain language for areas Biscay, Finisterre, Trafalgar, Norwegian Sea and Denmark Strait, also the Northern, Central and Southern sections of the region of the North Atlantic from 35°N to 71°N, between 15°W and 40°W. These sub-divisions are as shown on the chart on page 52. The forecasts are based on the fully analysed charts for midnight and 1200 UTC, supplemented as necessary by observations for 0600 and 1800 UTC that are to hand at the time of issue. The forecasts cover a period of 24 hours from the time of issue.

Transmission of Part 4, Analysis, in Analysis Code (I.A.C. FLEET) FM 46-IV, has been suspended from the Atlantic Weather Bulletin but as some National Meteorological Services continue to broadcast in I.A.C. Fleet Code, examples of the analysis are included in this publication.

Example of an Atlantic Weather Bulletin

From 'Bracknell Weather' to 'All Ships'.

Part 1. Storm Warnings

Storm warnings. At 190600 UTC complex low Bailey 978 moving slowly east-north-east. Winds expected to reach storm 10 at times locally at first in south-west quadrant between about 100 and 400 miles from centre.

Part 2. Synopsis of Weather Conditions

At 190001 UTC, complex low Bailey 978 moving slowly east-north-east. High 100 miles west of Azores 1037 moving slowly south-west declining a little. Low about 100 miles south-east of Newfoundland 1013 moving rather quickly east, expected 1020 by 200001 UTC.

---

* For details about the use of weather bulletins for shipping, see *Meteorology for Mariners*.

Part 3. Forecasts

Forecast for areas Biscay, Trafalgar, Finisterre, Sole and from 35 to 65 north between 15 and 40 west, and Norwegian Sea and Denmark Strait, for next 24 hours. (See map on page 52.)

Biscay, Finisterre — Mainly north 4 in south, otherwise west 4 or 5 veering north-west and increasing 6 or 7 in north. Occasional rain. Moderate with fog patches becoming good in north.

Trafalgar — North or north-east 4 or 5. Mainly fair. Moderate or good.

Sole — West or north-west 5 increasing 6 to gale 8. Showers. Good.

East Northern Section — In north, northerly 6 to gale 8 at first. In south, north-westerly 7 to severe gale 9 locally storm 10 at first. Thundery showers. Mainly good.

West Northern Section — In north, northerly backing westerly 4. In south, westerly 6 locally gale 8 at first. Occasional rain. Good, locally poor.

East Central Section — In north, westerly 6 locally gale 8 severe gale 9 at first. In south, north-west 5 becoming variable 4. Thundery showers in north. Occasional rain in south. Mainly good becoming moderate locally, poor in south.

West Central Section — In north-east, westerly 6 locally gale 8 decreasing 5. In north-west, south-westerly 4 or 5. In south, variable 4 locally 6 for a time. Showers in north, occasional rain in south. Good in north, moderate with fog patches in south.

East Southern Section — In north, north-west backing westerly 4 or 5. In south, north-east 4. Occasional rain in north. Good, but moderate in north.

West Southern Section — In north, mainly south-westerly 4 or 5 locally 6. In south, variable 3 or 4. Occasional rain in north. Moderate with fog patches in north, otherwise good.

Norwegian Sea and Denmark Strait — North-east 4 becoming 5 to 6 in east and locally 7 in south.

Global Maritime Distress and Safety System Broadcasts

High seas forecasts and warnings for METAREA I (see map on page 52) are broadcast from the BT INMARSAT station at Goonhilly, Cornwall, through the Atlantic Ocean Region East satellite (AOR(E)). Warnings are also broadcast through AOR(W) as soon as possible after receipt. The forecasts are forwarded via the International SafetyNET Service of INMARSAT Standard-C as Enhanced Group Call messages within the GMDSS.

They are broadcast in four similar parts and at approximately the same times as the HF W/T North Atlantic Bulletin, but the area forecasts cover the following areas: Sole, Shannon, Rockall, Bailey, Faeroes, South-east Iceland, Northern and Central Sections subdivided as necessary, and Norwegian Sea and Denmark Strait.

Storm warnings are included in the first section of the high seas weather bulletin issued by The Met. Office for broadcast via GMDSS arrangements at 0930 and 2130 UTC. Amended and new warnings issued at other times are broadcast as and when issued.

Working charts for use with the Atlantic Weather Bulletin*

The most suitable working chart for use with the Atlantic Weather Bulletin for shipping is obsolete Metform 1258A*, North Atlantic Plotting Chart. Copies of this chart are supplied to all ships of the United Kingdom Voluntary Observing Fleet regularly co-operating with The Met. Office. They may also be obtained from the Marine Division of The Met. Office, Met O(OM), Scott Building, Eastern Road, Bracknell, Berks RG12 2PW, telephone 01344 855656, facsimile 01344 855921.

Example Analysis

The following example in I.A.C. Fleet Code is plotted on Metform 1258A* in conjunction with the Atlantic Weather Bulletin. (See Section 5.)

```
10001 33300 01906
99900 84117 47468 80720 80180 59088 61126 00717
      85523 44578 85237 40328 10000
      83/// 54096 56078 60085 88010 63506
99911 66457 31818 32777 35726 37636 41555 45488 47468
      66250 47468 48437 47396 45255
      66450 45225 43156 44117 47046 51000 55020
      66650 55020 59003 63055 64105 62165
      66250 55020 52042 48062
      66450 42778 47655 53575
      66650 53575 56556 61577 63635
      66457 56101 60063 64040 68025 70105
99922 44984 58067 62087 63135 61156 58067
      44992 56078 59003 64045 66127 62217 58155 56078
      44000 55087 55020 60063 64040 69136 63306 57205 55087
      44008 70295 66438 60437 55205 52106 51000 52042 54101
      44016 30715 37636 40747 48605 57455 52206 47046 48050 34000
            33055 30085
      44024 30565 32578 42506 47396 52435 53406 49206 44117 42098
            30225
      44032 45306 42238 35306 33425 36468 45306
      44000 58685 59656 61665
      44008 55715 53628 56556 60535 64615
      44016 50745 48718 44775
      44020 30617 41555 42645 46606 48555 45488 48437 52477
            55426 55365 45076 45005
                  19191
```

---

* Metform 1258A is suitable for plotting practice but being out of print the meteorological station numbers are not updated.

Drawing the Isobars

Isobars are lines drawn on the chart through places having the same barometric pressure. To complete the chart, the best procedure to adopt is as follows:

(a) Mark in the positions of the centres of high and low pressure, checking to see that they agree reasonably well with the information already on the chart.
(b) Plot the positions delineating the fronts and join them with smooth lines. These lines denote the boundaries of the different air masses. It is usual on working charts to use red pencil for a warm front, blue pencil for a cold front, purple for an occlusion and other conventions as listed in the lower table opposite. Alternatively, but with considerably more trouble, one can plot with one colour using the symbols for printed charts shown in the same table.
(c) Next draw each isobar as given, first by plotting the several positions and then joining them with a smooth line. Additional isobars are inserted to make an interval of 4 mb between adjacent isobars. In drawing these isobars, remember that they are bent sharply at a front to accord with the experience that at a front the wind usually changes direction suddenly. The point of the discontinuity always protrudes from low pressure to high pressure. This agrees with the rule that the wind backs before the arrival of a front and veers suddenly at its passage. Within the warm sector of a 'Low', that is between the warm and cold fronts, the isobars are approximately straight lines.

Movement of pressure systems and fronts

The Analysis also contains information on the movement of the pressure systems and fronts. This information is generally based either on the past behaviour of the different systems or on theoretical considerations. For example, a depression, in general, moves parallel to the direction of the isobars within the warm sector and its speed is approximately equal to that of the geostrophic wind in the warm sector. The Geostrophic Scale on the working chart (Metform 1258A) may be used for obtaining geostrophic wind speed. To obtain a rough estimate of the surface wind speed over the sea, the wind speed obtained by using this scale must be reduced by one-third.

Beaufort notation to indicate the state of the weather

Weather	Beaufort letter	Weather	Beaufort letter
Blue sky (0–2/8 clouded)	b	Overcast sky. (The whole sky covered with unbroken cloud)	o
Sky partly clouded (3–5/8 clouded)	bc		
Cloudy (6–8/8 clouded)	c	Squally weather	q
Drizzle	d	Rain	r
Wet air (without precipitation)	e	Sleet (i.e. rain and snow together)	rs
Fog	f	Snow	s
Gale	g	Thunder	t
Hail	h	Thunderstorm with rain or snow	tlr or tls
Precipitation in sight of ship or station	jp	Ugly threatening sky	u
Line squall	kq	Unusual visibility	v
Storm of drifting snow	ks	Dew	w
Sandstorm or duststorm	kz	Hoar frost	x
Lightning	l	Dry air	y
Mist	m	Dust haze	z

Symbols for fronts

Fronts or boundaries between masses of air of different origin are indicated on a weather map, wherever their characteristics are well pronounced, in the following way:

Code figure	F_t Type of front	Used on printed charts	Used on working charts
0	Quasi-stationary front		Alternate red and blue lines joined together to form a continuous line
1	Quasi-stationary front above the surface		Alternate red and blue broken line
2	Warm front		Continuous red line
3	Warm front above the surface		Broken red line
4	Cold front		Continuous blue line
5	Cold front above the surface		Broken blue line
6	Occlusion		Continuous purple line
7	Instability line		Plotted in black
8	Intertropical front		Alternate red and green broken line
9	Convergence line		Solid orange line

Note. The symbols used on printed charts are placed on the side of the line towards which the front is moving.

SECTION 5
THE INTERNATIONAL ANALYSIS CODE
FM46-IV (I.A.C. FLEET)

The International Analysis Code is used by National Meteorological Stations to issue the results of detailed synoptic analyses in an abbreviated form suitable for rapid transmission. It is intended to assist the recipient to plot and interpret the weather map for a specified time. The groups in brackets are optional, to be used at the discretion of each service.

Analysis:

    10001   33388* $0YYG_cG_c$ or

Forecast:

    65556   33388* $0YYG_cG_c$ $000G_pG_p$

Pressure systems:

    99900
$$\begin{array}{l} 8P_tP_cPP \\ 000g_pg_p \\ \text{and/or} \\ 000g_pg_p \end{array} \quad \begin{array}{l} \\ 9P_tP_cPP \\ \\ 7P_tP_cPP \end{array} \quad \begin{array}{l} QL_aL_aL_oL_o \\ QL_aL_aL_oL_o \\ \\ QL_aL_aL_oL_o \end{array} \quad \begin{array}{l} (QL_aL_aL_oL_o) \\ (QL_aL_aL_oL_o) \\ \\ (QL_aL_aL_oL_o) \end{array} \quad \begin{array}{l} md_sd_sf_sf_s \\ \\ \\ md_sd_sf_sf_s \end{array}$$

Frontal systems:

    99911
$$\begin{array}{l} 66F_tF_iF_c \\ 000g_pg_p \\ \text{and/or} \\ 000g_pg_p \end{array} \quad \begin{array}{l} \\ 69F_tF_iF_c \\ \\ 67F_tF_iF_c \end{array} \quad \begin{array}{l} QL_aL_aL_oL_o \\ QL_aL_aL_oL_o \\ \\ QL_aL_aL_oL_o \end{array} \quad \begin{array}{l} (QL_aL_aL_oL_o) \\ (QL_aL_aL_oL_o) \\ \\ \ldots \end{array} \quad \begin{array}{l} md_sd_sf_sf_s \\ \\ \\ md_sd_sf_sf_s \end{array}$$

Isobars section:

    99922
    44PPP $QL_aL_aL_oL_o$ $(QL_aL_aL_oL_o)$ . . . .

Weather area section:

    99944
    $987w_sw_s$ $QL_aL_aL_oL_o$ $(QL_aL_aL_oL_o)$ . . . . $md_sd_sf_sf_s$

Tropical section:

    99955
    $(55T_tT_iT_c)$ (555PP) $QL_aL_aL_oL_o$ $(QL_aL_aL_oL_o)$ . . . . $md_sd_sf_sf_s$

---

* When the second group is 33388 the position group is in the form $QL_aL_aL_oL_o$. Some meteorological services may report position to the nearest half degree in the form of $L_aL_aL_oL_ok$. In such cases the second group will be reported as 33300, 33311 or 33322 where:
        33300 = Northern Hemisphere
        33311 = Southern Hemisphere
        33322 = Equatorial region
In the equatorial region latitudes from 0°–30°S are indicated by subtraction from 100 e.g. 13°S, $L_aL_a = 87$.

Wave or sea temperature section:

$$\begin{gathered}
88800 \\
77e_2uu \quad (9d_wd_wP_wP_w) \quad QL_aL_aL_oL_o \quad (9d_wd_wP_wP_w) \ldots md_sd_sf_sf_s
\end{gathered}$$

$$\left[\begin{array}{llll}
QL_aL_aL_oL_o & \ldots\ldots & (9d_wd_wP_wP_w) & QL_aL_aL_oL_o & (00C_100) \\
000g_pg_p & 79e_2uu & (9d_wd_wP_wP_w) & QL_aL_aL_oL_o & (9d_wd_wP_wP_w) \\
QL_aL_aL_oL_o & \ldots\ldots & (9d_wd_wP_wP_w) & QL_aL_aL_oL_o & (00C_100) \\
\text{and/or} & & & & \\
000g_pg_p & 76e_2uu & (9d_wd_wP_wP_w) & QL_aL_aL_oL_o & (9d_wd_wP_wP_w) \\
QL_aL_aL_oL_o & \ldots\ldots & (9d_wd_wP_wP_w) & QL_aL_aL_oL_o & (00C_100) \\
77744 & \ldots\ldots & \text{Plain Language} & \ldots\ldots & 44777
\end{array}\right]$$

$$19191$$

Indicator groups 10001   65556   33388*

- 10001   Indicator group signifying "Analysis following".
- 65556   Indicator group signifying "Prognostic (Forecast) Analysis following".
- 33388* Indicator group for position fixing system.

$0YYG_cG_c$

- 0   Indicator figure.
- YY Day of the month.
- G_cG_c Synoptic hour, in UTC, of observation of data from which chart is prepared.

$000G_pG_p$

- 000   Indicator figures.
- G_pG_p Number of hours to be added to G_cG_c (chart time) to obtain time to which the prognosis refers.

$000g_pg_p$

- 000   Indicator figures.
- g_pg_p Number of hours to be added to, or subtracted from, G_cG_c to give supplementary information.

99900 Pressure system section

$8P_tP_cPP$ $9P_tP_cPP$ $7P_tP_cPP$

- 8   Position as for G_cG_c.
- 9   Position as for $G_cG_c - g_pg_p$.
- 7   Position as for $G_cG_c + g_pg_p$.
- P_t Type of pressure system (Table 30).
- P_c Character of pressure system (Table 30).
- PP Pressure at centre of system in millibars (omitting the thousands and hundreds figures).

$QL_aL_aL_oL_o$ Position group*.

- Q   Octant of the globe (Table 28).

$md_sd_sf_sf_s$ Movement group

- m   Movement indicator figure (Table 31).
- d_sd_s Direction, in tens of degrees, towards which system is moving (01–36; 00 = stationary, 99 = unknown).
- f_sf_s Speed, in knots, of moving system 99 = unknown.

---

* See footnote on page 36

99911 Frontal system section
66$F_tF_iF_c$ 69$F_tF_iF_c$ 67$F_tF_iF_c$

66	Position as for G_cG_c.
69	Position as for $G_cG_c - g_pg_p$.
67	Position as for $G_cG_c + g_pg_p$.
F_t	Type of front (Table 32).
F_i	Intensity of front (Table 32).
F_c	Character of front (Table 32).
Q$L_aL_aL_oL_o$	Position group*.
m$d_sd_sf_sf_s$	Movement group (as under Pressure systems).

99922 Isobars section
44PPP   Q$L_aL_aL_oL_o$

44	Indicator figures signifying isobars follow.
PPP	Pressure in millibars (omitting the thousands figure).
Q$L_aL_aL_oL_o$	Position group*.

99944 Significant weather data section
987w_sw_s

987	Indicator figures signifying significant weather data follows.
w_sw_s	Significant weather (Table 33).
Q$L_aL_aL_oL_o$	Position group*.

99955 Tropical systems section
55$T_tT_iT_c$

55	Indicator figures signifying tropical system follows.
T_t	Tropical circulation type (Table 34).
T_i	Tropical system intensity (Table 34).
T_c	Tropical system characteristic (Table 34).
555PP	Optional group giving pressure at centre of system.
555	Indicator figures.
PP	Pressure to nearest millibar (omitting the thousands and hundreds figures).
Q$L_aL_aL_oL_o$	Position group*.
m$d_sd_sf_sf_s$	Movement group (as under Pressure systems).

88800 Wave and/or Sea temperature isopleth section
77$e_2$uu 79$e_2$uu 76$e_2$uu

77	Position as for G_cG_c.
79	Position as for $G_cG_c - g_pg_p$.
76	Position as for $G_cG_c + g_pg_p$.
$e_2$uu	Type of isopleth and units used (Table 35).
9$d_wd_wP_wP_w$	Wave group.
d_wd_w	Direction in tens of degrees from which waves are coming (Table 8, p.11).
P_wP_w	Period of waves in seconds.
Q$L_aL_aL_oL_o$	Position group*.
00$C_1$00	Confidence group *(Table 36).

---

* See footnote on page 36

Plain language section 77744 44777

77744   Indicator group signifying "Plain language following".
44777   Indicator group signifying end of plain language.

19191: End of message group

Explanation

In spite of the large number of symbols and specifications used, the Analysis Code form is inherently simple if the following facts are remembered:

(a) If the first group is 10001 this signifies that an Analysis follows. If the first group is 65556 this signifies that a Prognostic (Forecast) Analysis follows.

(b) If the second group is 33300, 33311 or 33322 instead of 33388 the positions are given to the nearest half degree.

(c) In sections 88800, 99900 and 99911 one or more alternative groups, introduced by a time group $000g_pg_p$, may be used when it is required to give greater detail of the movement and characteristic of any particular system.

In the section 99900 of an analysis message the groups $000g_pg_p$ $9P_tP_cPP$ $QL_aL_aL_oL_o$ give details of pressure systems at g_pg_p hours *prior* to G_cG_c. When the groups $000g_pg_p$ $9P_tP_cPP$ $QL_aL_aL_oL_o$ are used the details given refer to pressure systems at g_pg_p hours *after* G_cG_c.

Similarly frontal systems, wave systems and sea temperature configuration can be treated in the same way. The same principle applies to prognosis messages, where g_pg_p is added to or subtracted from the time $G_cG_c + G_pG_p$.

(d) A group beginning with 8, 9 or 7 gives data relating to a pressure system, e.g. a depression. This group is followed by one or more position groups as necessary. Similar information is given for each pressure system included in the analysis, i.e. there may be several groups beginning with 8, 9 or 7, each followed by one or more position groups.

(e) A group beginning with 66, 69 or 67 gives data relating to a front, the succeeding groups specifying its position. There are generally several groups beginning with 66, 69 or 67, each followed by a series of position groups.

(f) The figures 44 are used to indicate data relating to isobars, the figures PPP in the group 44PPP giving the numerical value of the isobar whose position is determined by the series of following position groups.

(g) Groups beginning with 55 are used only for tropical regions. When these groups are included, groups giving frontal data. i.e. those beginning with 66, 69 or 67, are usually omitted.

(h) The figures 987 are used to indicate data relating to weather.

(i) A group beginning with 77, 79 or 76 gives detail of waves and/or sea temperature isopleths.

(j) The groups contained by the indicator groups 77744 and 44777 are in plain language.

(k) The group $md_sd_sf_sf_s$, may be included after the data for each pressure system or front to indicate its direction and speed of movement.

(l) When it is necessary to send a correction to the analysis or prognosis the correction message commences with the groups 11133 $0YYG_cG_c$. The corrections follow, preceded by the appropriate indicators (8....., 66...., 44...., etc.), and the message ends with the 19191 group.

ANALYSIS SPECIFICATION TABLES

Table 28

Q = Octant of the globe

	Longitude				Longitude	
0	0°–90°W	North Latitude		5	0°–90°W	South Latitude
1	90°W–180°			6	90°W–180°	
2	180°–90°E			7	180°–90°E	
3	90°E–0°			8	90°E–0°	

Table 29

k = Half-degree position figure

0	Take $L_aL_aL_oL_o$ as sent		5	Take $L_aL_aL_oL_o$ as sent	
1	Add ½ degree to L_aL_a	0°–99°E	6	Add ½ degree to L_aL_a	0°–99°W
2	Add ½ degree to L_oL_o	or	7	Add ½ degree to L_oL_o	or
3	Add ½ degree to L_aL_a and L_oL_o	100°W–180°	8	Add ½ degree to L_aL_a and L_oL_o	100°E–180°
4	Whole degrees		9	Whole degrees	

Note. Half degree position figure "k": L_aL_a and L_oL_o are the latitude and longitude in whole degrees (for longitude west of 100°W and east of 100°E, the initial 1 is omitted). For equatorial regions, latitudes from 0° to 30°S are subtracted from 100 (e.g. L_aL_a 87 = 13°S; L_aL_a 71 = 29°S). When k = 4 or 9, the values of L_aL_a and L_oL_o are accurate to the nearest whole degree only; for all other values of k the accuracy is to the nearest ½ degree.

Table 30

P_t, P_c = Pressure systems

	P_t Type of pressure system	P_c Character of pressure system	
0	Complex low	No specification	0
1	Low	Low filling or high weakening	1
2	Secondary	Little change	2
3	Trough*	Low deepening or high intensifying	3
4	Wave	Complex	4
5	High	Forming or existence suspected (cyclogenesis or anticyclogenesis)	5
6	Area of uniform pressure	Filling or weakening but not disappearing	6
7	Ridge	General rise of pressure	7
8	Col	General fall of pressure	8
9	Tropical storm	Position doubtful	9

* A trough line is shown on charts as a continuous black line.

Table 31

m = Movement indicator figure

0	No specification	5	Curving to left
1	Stationary	6	Recurving
2	Little change	7	Accelerating
3	Becoming stationary	8	Curving to right
4	Retarding	9	Expected to recurve

Table 32

F_t, F_i, F_c = Frontal systems

	F_t Type of front	F_i Intensity of front	F_c Character of front	
0	Quasi-stationary front	No specification	No specification	0
1	Quasi-stationary front above the surface	Weak, decreasing (including frontolysis)	Frontal activity area decreasing	1
2	Warm front	Weak, little or no change	Frontal activity area, little change	2
3	Warm front above the surface	Weak, increasing (including frontogenesis)	Frontal activity area increasing	3
4	Cold front	Moderate, decreasing	Intertropical	4
5	Cold front above the surface	Moderate, little or no change	Forming, or existence suspected	5
6	Occlusion	Moderate, increasing	Quasi-stationary	6
7	Instability line*	Strong, decreasing	With waves	7
8	Intertropical front	Strong, little or no change	Diffuse	8
9	Convergence line	Strong, increasing	Position doubtful	9

* An addition in plain language may be made, when this is considered necessary to emphasise the existence of a line squall.

Table 33

$w_s w_s$ = Significant weather

00	Area of heavy swell	55	Area of gales (Beaufort 8 or more)
11	Area of strong winds (6 and 7 Beaufort)	66	Area of continuous precipitation
22	Area of middle cloud	77	Area of squally weather
33	Area of low cloud	88	Area of heavy showers
44	Area of poor visibility	99	Area of thunderstorms

Table 34

T_t, T_i, T_c = Tropical systems

	T_t Tropical circulation type	T_i Tropical system intensity when $T_t = 0$ to 8	T_i Tropical system intensity when $T_t = 9$	T_c Tropical system characteristic	
0	Intertropical convergence zone	No specification	Force 10	No specification	0
1	Shear line	Weak, decreasing	Force 11	Diffuse	1
2	Line or zone of convergence	Weak, little or no change	Force 12 64–71 kn	Sharply defined	2
3	Axis of doldrum belt	Weak, increasing	72–80 kn	Quasi-stationary	3
4	Trough in westerlies	Moderate, decreasing	81 kn or over	Existence certain	4
5	Trough in easterlies	Moderate, little or no change	Force 5	Existence uncertain	5
6	Low area	Moderate, increasing	Force 6	Formation suspected	6
7	Surge line	Strong, decreasing	Force 7	Position certain	7
8	Line or zone of divergence	Strong, little or no change	Force 8	Position uncertain	8
9	Tropical cyclonic circulation	Strong, increasing	Force 9	Movement doubtful	9

Table 35

e_2 = Type of isopleth (and unit of isopleth values uu)

0	Sea wave height isopleth, uu in metres.
1	Swell wave height isopleth, uu in metres.
2	Wave height isopleth (wave type undetermined), uu in metres.
3	Wave direction isopleth, uu in tens of degrees.
4	Wave period isopleth, uu in seconds.
5	Reserved.
6	Reserved.
7	Reserved.
8	Reserved.
9	Sea temperature isopleth, uu in whole degrees Celsius.

Table 36

C_1 = Confidence figure (used in group $00C_100$)

0	No specification.		5	Uncertain.
2	With confidence.		8	Very doubtful.

Plotting observations broadcast in weather bulletins

Working charts can be obtained from most Meteorological Services which issue weather bulletins for ships. The United Kingdom Met. Office supplies Metform 1258A* for the North Atlantic and has available Admiralty charts for other areas for free issue to Voluntary Observing Ships.

It will be noticed that on the weather chart on p.60 the index numbers of the stations used in the bulletin are printed alongside the station circles, the block numbers of the different land areas being printed in large figures within a square 'box'. Details are given on p.46.

The content of each group in the coded reports from ships and land stations is indicated in Sections 1 and 2 (pp.3–7), the corresponding plotting symbols, where they are needed, being found on p.45.

The various elements contained in the message are plotted around the station circle, in figures and symbols, in the positions shown in the example below.

Decoded message

Station Model

$TTT\ \ PPPP$
$VVww\ \textcircled{N}$
$\ \ \ \ \ \ W_1W_2$
$P_wP_wH_wH_w\ D_sv_s$
$d_{w1}d_{w1}P_{w1}P_{w1}H_{w1}H_{w1}$

VV	=	Visibility (2.2 n.miles = 96 in code)
N	=	Cloud amount (8/8)
dd	=	Wind direction (North)
ff	=	Wind force (40 knots)
TTT	=	Air temperature (06.4)
PPPP	=	Barometric pressure (992.4 mb)
ww	=	Present weather (continuous moderate rain)
W_1W_2	=	Past weather (showers, cloud cover ½ or more)
D_s	=	Ship's course (East)
v_s	=	Ship's speed 18 knots, code figure 4
P_wP_w	=	Period of sea waves (06 seconds)
H_wH_w	=	Height of sea waves (5½ metres = 11 in message)
$d_{w1}d_{w1}$	=	Direction from which swell waves are coming (230 degrees, 23 in code)
$P_{w1}P_{w1}$	=	Period of swell waves (12 seconds)
$H_{w1}H_{w1}$	=	Height of swell waves (4 metres = 08 in message)

As Plotted

064 9924
96 •• ●
    0611 ▽
       → 4
    1208

The direction of the wind is represented by an arrow flying with the wind, and the speed by the number of feathers or pennants on the arrow. Full feathers each represent 10 knots and a half feather 5 knots. A pennant signifies 50 knots. (See table on page 45.)

Barometric pressure, air temperature, visibility, period and height of waves and/or swell are plotted as received. Total cloud amount, present weather and past weather are plotted using the appropriate symbols. (See p.45.) The symbol for $d_{w1}d_{w1}$ is shown, the arrow head pointing with the direction of swell.

---

* See footnote on page 33.

Ships' Reports. (See chart opposite page 36)

BBXX YYGGi$_w$
99L$_a$L$_a$L$_a$ Q$_c$L$_o$L$_o$L$_o$L$_o$ i$_{Ri}$hVV Nddff 1s$_n$TTT 4PPPP 7wwW$_1$W$_2$ 222D$_s$V$_s$
2P$_w$P$_w$H$_w$H$_w$ 3d$_{w1}$d$_{w1}$// 4P$_{w1}$P$_{w1}$H$_{w1}$H$_{w1}$

BBXX19061
99527 70355 42598 82408 10081 40264
22200 20302 326 // 41208
99415 70276 41497 83005 10179 40335 75055
22214 20101 328 // 40503
99484 70392 41397 81108 10128 40239 76162
22213 20302 312 // 40604
BBXX19064
99460 70179 42597 73317 10142 40264
22211 20705 336 // 41509
99528 70222 41498 82835 10108 40192 78082
22272 20510 325 // 41009
BBXX19063
99570 70205 41498 52741 10089 40025 71588
22261 21014 324 // 41010
99408 70183 41194 83010 10183 40290 75211
22233 20302 329 // 40604
99395 70398 42598 22004 10209 40352
22223 20201
99642 70365 42397 80212 11009 44051
22211 20304
99395 70113 42498 80115 10172 40230
22243 20404 335 // 40505
99470 70072 41297 82910 10138 40188 75222
22213 20202 333 // 40706

Station Reports. (See chart opposite page 36)

AAXX YYGGi$_w$
IIiii i$_{Ri}$hVV Nddff 1s$_n$TTT 4PPPP 7wwW$_1$W$_2$
AAXX19064
01203 32680 61305 10119 49997
03026 11560 81823 10102 49853 76082
03804 32481 72916 10129 40138
03953 31565 72718 10118 40111 70262
03976 11470 72725 10119 40027 72588
04030 32684 31804 10082 49927
04390 32980 12618 10076 40078
06011 11397 81608 10093 49851 76166
07110 11256 83010 10134 40152 70255
07510 11360 72804 10154 40201 72052
08001 11562 80000 10147 40233 72582
08505 32580 20209 10174 40332
71600 21008 92005 10069 40210 74544
71818 32974 02015 10092 40093
78016 32561 72205 10243 40176

BBXX = A bulletin of ship reports follows.
AAXX = A bulletin of station reports follows.

ff	Symbol	ff	Symbol
kt		kt	
Calm	◎	33–37	
1–2		38–42	
3–7		43–47	
8–12		48–52	
13–17		55–57	
18–22		58–62	
23–27		63–67	
28–32		68–72	
Wind direction given but speed missing			
Wind direction variable			

In plotting, the wind arrow should be entered first to allow for any possible displacement of the other elements around the station circle. To convert wind speeds reported in metres per second, double ff and then plot in the normal way; i.e. 10 m/sec ≃ 20 knots.

Symbols used for plotting on working charts

45

International meteorological codes for reports from land stations

FM12-X Ext. Full message

$YYGGi_w$ (II)iii i_Ri_xhVV Nddff (00fff) $1s_nTTT$ $2s_nT_dT_dT_d$ ($3P_oP_oP_o$)
4PPPP 5appp ($6RRRt_R$) $7wwW_1W_2$ $8N_hC_LC_MC_H$ (9hh///)

Reduced message

This is the code used in radio weather bulletins for shipping.
$YYGGi_w$ (II)iii i_Ri_xhVV Nddff (00fff) $1s_nTTT$ 4PPPP $7wwW_1W_2$

Group IIiii

    II   Regional block number.
    iii   Station number.

All meteorological stations ashore have been allotted 5-figure index numbers which are reported in the coded form IIiii where II denotes an area, and iii a station within that area. The first group from individual stations may consist either of the 5-figure index number IIiii or the 3-figure station number iii. Most countries in the European Region, including the United Kingdom, have adopted the 3-figure station number and to enable such stations to be identified in collective messages a sequence of reports from stations in the same block is preceded by a group 999II for the purpose of identifying the block. It is not necessary to use the 999III group in collective messages when the individual reports begin in the form IIiii.

The general scheme of assignment of block numbers to Geographical Regions is as follows:

Europe and Asia	00 to 49
Africa	60 to 69
North and Central America	70 to 79
South America and Antarctic	80 to 89
Australia and Pacific Ocean	90 to 99

The individual stations within each block are listed in numerical order, together with details of their latitude, longitude and height, in *Admiralty List of Radio Signals*, Vol. 4.

Block number 03 has been assigned to the British Isles. Station numbers 03000 to 03899 apply to Great Britain and 03900 to 03999 to Northern Ireland and the Irish Republic.

SECTION 6
FORECASTS AND WEATHER INFORMATION AVAILABLE TO SHIPPING

Full details of the Atlantic Weather Bulletin and the International Analysis Code are given in Sections 4 and 5. The International Maritime Forecast (MAFOR) code is included in this Section. Details of forecasts and other weather information broadcast in plain language from UK and other countries are given in *Admiralty List of Radio Signals*, Vol. 3.

Most meteorological services provide bulletins in plain language (first in the language of the country concerned, then in English), supplemented in some cases by reports in one or more of the following code forms:

FM 12-X Ext. (see page 46)
FM 13-X (see page 3)
FM 45-IV (see *Admiralty List of Radio Signals*, Vol. 3).
FM 46-IV (see page 36).

Details of facsimile weather charts available to shipping are published in the *Admiralty List of Radio Signals*, Vol. 3. Symbols used on these charts are shown on pages 54, 55.

Some countries, which do not issue plain language bulletins in English, provide bulletins in the "MAFOR" code, FM 61-IV.

Maritime Forecast Code, FM 61-IV (MAFOR)

This code form, which has been adopted for international use by the World Meteorological Organization, may be used by certain meteorological services in weather bulletins for shipping when there are difficulties with respect to the broadcast of the forecast in English.

The specific value of any of the elements given in the forecasts should be understood to be necessarily approximate and the value of the element in question should accordingly be interpreted as representing the most probable mean of a range of values which the element may assume during the period of the forecast concerned and over the area concerned.

FM 61-IV

MAFOR $YYG_1G_1/$ $0AAAa_m$ $1GDF_mW_m$ $(2VST_xT_n)$ $(3D_KP_wH_wH_w)$

Group MAFOR

This code name is used as a prefix to the message, indicating that it is a forecast for shipping. If several of these messages are grouped in a single broadcast the prefix will appear only at the beginning of the collective message.

Group $YYG_1G_1/$

This group shall be used to give the UTC time and date of the beginning of the period for which the whole forecast or set of forecasts is valid. It need not be repeated if forecasts for several areas (AAA) are given in the one message.

Group 0AAAa$_m$

0	Indicator figure for this group.
AAA	Indicator for maritime area (for details see *Admiralty List of Radio Signals,* Vol. 3, under appropriate country).
a$_m$	An indicator for the portion of the maritime area (Table 37).

Note. The group 0AAAa$_m$ may be replaced by the geographical name for the forecast region.

Group 1GDF$_m$W$_m$

1	Indicator figure for this group.
G	Period of time covered by forecast (Table 38).
D	Direction of surface wind (Table 39).
F$_m$	Force of the surface wind (Table 40).
W$_m$	Forecast weather (Table 41).

Group (2VST$_x$T$_n$) *(This group is optional)*

2	Indicator figure for this group.
V	Visibility at surface (Table 42).
S	State of sea (Table 43).
T$_x$	Maximum air temperature (Table 44).
T$_n$	Minimum air temperature (Table 44).

Group (3D$_K$P$_w$H$_w$H$_w$) *(This group is optional)*

3	Indicator figure for this group.
D$_K$	Direction of swell (Table 39).
P$_w$	Period of waves (Table 45).
H$_w$H$_w$	Height of waves in units of ½ metres.

Note. The set of groups 1GDF$_m$W$_m$ (2VST$_x$T$_n$) (3D$_K$P$_w$H$_w$H$_w$) can be repeated as many times as necessary to describe the changes in the meteorological conditions forecast in a given area, due attention being given to the need for strict economy in the number of groups used. The first group 1GDF$_m$W$_m$ in which G = 1 to 8, and the following optional group(s), if used, then refer to the forecast weather commencing at the time given in the group YYG$_1$G$_1$/ and continuing through the period indicated by G. Subsequent groups 1GDF$_m$W$_m$ (G = 1 to 8) give the period of the time that the described weather is forecast to persist, commencing at the end of the period covered by the preceding group 1GDF$_m$W$_m$ (G = 1 to 8). Any set 1GDF$_m$W$_m$ (2VST$_x$T$_n$) (3D$_K$P$_w$H$_w$H$_w$) (G = 1 to 8) may be followed by a group 1GDF$_m$W$_m$ (G = 9) describing a phenomenon which is forecast to occur occasionally in the same period.

Tables for use with 'MAFOR' code

Table 37

a$_m$ = Indicator for the portion of the maritime area

Code figure		Code figure	
0	Whole of the area AAA	5	SW quadrant of the area AAA
1	NE quadrant of the area AAA	6	Western half of the area AAA
2	Eastern half of the area AAA	7	NW quadrant of the area AAA
3	SE quadrant of the area AAA	8	Northern half of the area AAA
4	Southern half of the area AAA	9	Rest of the area AAA

Table 38

G = Period of time covered by forecast

Code figure	
0	Synopsis of meteorological conditions in the forecast area at the time of the beginning of forecast period.
1	Forecast valid for 3 hours
2	Forecast valid for 6 hours
3	Forecast valid for 9 hours
4	Forecast valid for 12 hours
5	Forecast valid for 18 hours
6	Forecast valid for 24 hours
7	Forecast valid for 48 hours
8	Forecast valid for 72 hours
9	Occasionally

Table 39

D = Direction of surface wind

D_K = Direction of swell

Code figure	Direction	Code figure	Direction
0	Calm	5	SW
1	NE	6	W
2	E	7	NW
3	SE	8	N
4	S	9	Variable (for D code) / Confused (for D_K code)

Table 40

F_m = Force of the surface wind

Code figure	Beaufort number	Code figure	Beaufort number
0	0–3	5	8
1	4	6	9
2	5	7	10
3	6	8	11
4	7	9	12

Table 41

W_m = Forecast weather

Code figure
- 0   Moderate or good visibility (greater than 5 km/3 n. miles).
- 1   Risk of accumulation of ice on superstructures (air temperature between 0 °C/32 °F and -5 °C/23 °F).
- 2   Strong risk of accumulation of ice on superstructures (air temperature below -5 °C/23 °F).
- 3   Mist (Visibility 1 km/⅝ n. miles to 5 km/3 n. miles).
- 4   Fog (Visibility less than 1 km/⅝ n. miles).
- 5   Drizzle.
- 6   Rain.
- 7   Snow or rain and snow.
- 8   Squally weather with or without showers.
- 9   Thunderstorms.

Table 42

V = Visibility at surface

Code figure
- 0   Less than 50 metres.
- 1   50–200 metres.
- 2   200–500 metres.
- 3   500–1,000 metres (approx. 500 m–⅝ n. miles).
- 4   1–2 km (approx. ⅝–1 n. miles).
- 5   2–4 km (approx. 1–2 n. miles).
- 6   4–10 km (approx. 2–6 n. miles).
- 7   10–20 km (approx. 6–12 n. miles).
- 8   20–50 km (approx. 12–30 n. miles).
- 9   50 km or more (30 n. miles or more).

Table 43

S = State of sea

Code figure	Descriptive terms	Height* in metres
0	Calm — glassy	0
1	Calm — rippled	0–0.1
2	Smooth — wavelets	0.1–0.5
3	Slight	0.5–1.25
4	Moderate	1.25–2.5
5	Rough	2.5–4
6	Very rough	4–6
7	High	6–9
8	Very high	9–14
9	Phenomenal	Over 14

* The average wave height as obtained from the larger well-formed waves of the wave system being observed.

Table 44

T_x = Maximum air temperature

T_n = Minimum air temperature

Code figure	°C	°F
0	Less than −10	Less than 14
1	−10 to −5	14 to 23
2	−5 to 1	23 to 30
3	About 0	About 32
4	1 to 5	34 to 41
5	5 to 10	41 to 50
6	10 to 20	50 to 68
7	20 to 30	68 to 86
8	greater than 30	greater than 86
9	Temperature not forecast	

Table 45

P_w = Period of waves

Code figure	Period in seconds	Code figure	Period in seconds
5	5 or less	0	10
6	6	1	11
7	7	2	12
8	8	3	13
9	9	4	14 or more
/	calm or period not determined		

Note. The period of the waves is the time between the passage of two successive wave crests past a fixed point. The average value of the wave period is reported, as obtained from the larger well-formed waves of the wave system being observed.

Chart showing the areas referred to in Weather Bulletins and Gale warnings broadcast by the BBC and BT coast radio stations

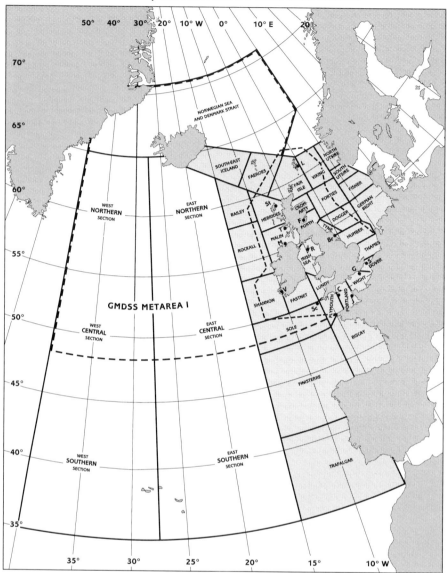

Stations whose latest reports are broadcast in the 5-minutes forecast

T = Tiree; **St** = Stornoway; **L** = Lerwick; **F** = Fife Ness; **Br** = Bridlington; **S** = Sandettie Light-Vessel Automatic; **G** = Greenwich Light-Vessel Automatic; **J** = Jersey; **C** = Channel Light-Vessel Automatic; **Sc** = Scilly Automatic; **V** = Valentia; **R** = Ronaldsway; **M** = Malin Head.

Boundaries of sea areas, used in weather forecasts transmitted by the BBC and BT coastal radio stations. The dashed line around the U.K. coast encloses GMDSS sea area A2. The North Atlantic areas, also Biscay, Trafalgar, Finisterre and Sole are included in the North Atlantic Bulletin. (See page 31). The dashed line across the Atlantic encloses GMDSS METAREA I, from latitude 48° 27′N to 71° 00′N. (See page 32).

Details of forecasts for the other areas are contained in *Admiralty List of Radio Signals*, Vol. 3. For easy reference, they are also contained in The Met. Office brochure, *Weather Services for Shipping*.

CONVERSION TABLE

feet/metres

(with reference to $H_w H_w$ = height of waves)

Feet	Metres	Code figure	Feet	Metres	Code figure
1	0.3	1	41	12.5	25
2	0.6	1	42	12.8	25
3	0.9	2	43	13.1	26
4	1.2	2	44	13.4	27
5	1.5	3	45	13.7	27
6	1.8	3	46	14.0	28
7	2.1	4	47	14.3	29
8	2.4	5	48	14.6	29
9	2.7	5	49	15.0	30
10	3.1	6	50	15.2	30
11	3.3	7	51	15.5	31
12	3.7	7	52	15.8	32
13	4.0	8	53	16.1	32
14	4.3	9	54	16.5	33
15	4.6	9	55	16.8	34
16	4.9	10	56	17.1	34
17	5.2	10	57	17.4	35
18	5.5	11	58	17.7	35
19	5.8	12	59	18.0	36
20	6.1	12	60	18.3	36
21	6.4	13	61	18.6	37
22	6.7	13	62	18.9	38
23	7.0	14	63	19.2	38
24	7.3	15	64	19.5	39
25	7.6	15	65	19.8	40
26	7.9	16	66	20.1	40
27	8.2	16	67	20.4	41
28	8.5	17	68	20.7	41
29	8.8	18	69	21.0	42
30	9.1	18	70	21.3	43
31	9.5	19	71	21.6	43
32	9.7	19	72	21.9	44
33	10.1	20	73	22.3	45
34	10.4	21	74	22.6	45
35	10.7	21	75	22.9	46
36	11.0	22	76	23.2	46
37	11.3	23	77	23.5	47
38	11.6	23	78	23.8	48
39	11.9	24	79	24.1	48
40	12.2	24	80	24.4	49

SYMBOLS USED ON RADIO-FACSIMILE WEATHER CHARTS

– FRONTAL AND PRESSURE-FEATURES SYMBOLS

Cold front at the surface		Quasi-stationary front above the surface	
Cold front above the surface			
Cold-front frontogenesis		Quasi-stationary front frontogenesis	
Cold-front frontolysis		Quasi-stationary front frontolysis	
Warm front at the surface		Instability line	
Warm front above the surface		Shear line	
Warm-front frontogenesis		Convergence line	
Warm-front frontolysis			
Occluded front at the surface		Intertropical convergence zone*	
Occluded front above the surface		Intertropical discontinuity	
Quasi-stationary front at the surface		Axis of trough	
		Axis of ridge	

* *Note:* The separation of the two lines gives a qualitative representation of the width of the zone; the hatched lines may be added to indicate areas of activity.

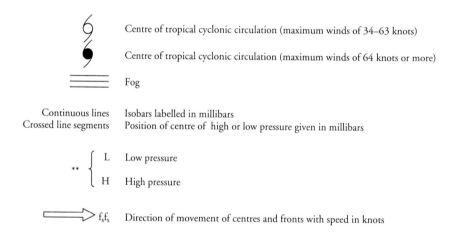

	Centre of tropical cyclonic circulation (maximum winds of 34–63 knots)
	Centre of tropical cyclonic circulation (maximum winds of 64 knots or more)
	Fog
Continuous lines	Isobars labelled in millibars
Crossed line segments	Position of centre of high or low pressure given in millibars
L	Low pressure
H	High pressure
$f_s f_s$	Direction of movement of centres and fronts with speed in knots

ICE ACCRETION

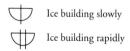

Ice building slowly

Ice building rapidly

** *Note:* The appropriate letter of the alphabet of the issue country may be used.

WAVE CHARTS

Continuous lines	Significant wind-wave height (sea), or composite wind-wave and swell height, where so drawn, labelled in metres
Dashed lines	Significant swell height, labelled in metres
MAX	Centre of maximum wave height
MIN	Centre of minimum wave height
⟶	Direction of sea waves
∿⟶	Direction of swell waves

SEA TEMPERATURE CHARTS

Continuous lines   Isotherms labelled in degrees Celsius

Note: Broken lines may be used to avoid confusion with other analysed parameters.

NEPHANALYSIS CHARTS

⌒⌒	Cumuliform cloud	⌒⌒	Apparent Cu con. or Cb	⟹	Direction of shear of Cirrus — from Cb anvil or other source
⟶	Cirriform cloud	///	Stratiform cloud	///	Wave clouds (mountain or transverse)
∿∿∿	Boundary of major cloud systems — fronts, vortices or other system dominating the scene viewed by the satellite	⟶ ⟶			Estimate location of jet stream
────	Definite boundary of more or less unorganized cloud masses	⤵	Vortex		
─ ─ ─	Indefinite boundary of more or less unorganized cloud masses	+	Heavy cloud	─	Thin cloud
⟵⟶	Striations	⟵- -⟶	Striations, tenuous		
⌒⌒	Cloud lines	⌒⌒ ⌒⌒	Building along the line		
⌒⌒ ⌒⌒	Cloud lines, tenuous-cloud form denoted by	{ ⌒⌒ ⌒⌒ /// ⟶ }			

Cloud cells	Size n.mile	Open spaces
1	0–30	6
2	30–60	7
3	60–90	8
4	90–120	6

CLOUD AMOUNT

Open (O) = Less than 20 per cent coverage	Mostly covered (MCO) = 50–80 per cent coverage
Mostly open (MOP) = 20–50 per cent coverage	Covered (C) = More than 80 per cent coverage

Note: Stippling is used to emphasize the area considered by the analyst to be of greater synoptic significance.

ALPHABETICAL LIST OF CODE SYMBOLS

Code symbols used only in decoding are distinguished by the number of the code at the end of the meaning. Where reference is to a code table the page number is in **bold type**.

Symbol	Meaning	Page
AAA	Indicator for maritime area (FM 61-IV)	48
A_1	WMO Region for offshore platform	4
a	Characteristic of barometric tendency	5, **14**
a_m	Indicator for the portion of the maritime area (FM 61-IV)	48, **48**
b_i	Ice of land origin	7, **25**
b_w	Sub-area of A_1	4
C_1	Confidence figure (FM 46-IV)	38, **42**
C_H	Type of High Cloud	5, **23**
C_L	Type of Low Cloud	5, **21**
C_M	Type of Medium Cloud	5, **22**
c_i	Concentration or arrangement of sea ice	7, **24**
D	Direction of surface wind (FM 61-IV)	48, **49**
D_K	Direction of swell (FM 61-IV)	48, **49**
D_i	Bearing of ice edge	7, **26**
D_s	Ship's course	6, **23**
dd	Wind direction in tens of degrees	4, **11**
$d_s d_s$	Direction towards which system or front is moving (FM 46-IV)	37
$d_w d_w$	Direction from which waves are coming (FM 46-IV)	38, **11**
$d_{w1} d_{w1}$	Direction from which first swell waves are coming	6, **11**
$d_{w2} d_{w2}$	Direction from which second swell waves are coming	6, **11**
$E_s E_s$	Thickness of ice accretion	7
c_2	Type of isopleth (FM 46-IV)	38, **42**
F_c	Character of front (FM 46-IV)	38, **41**
F_i	Intensity of front (FM 46-IV)	38, **41**
F_m	Beaufort force of wind (FM 61-IV)	48, **49**
F_t	Type of front (FM 46-IV)	35, 38, **41**
ff	Wind speed	4, **12**, **45**
fff	Wind speed, 99 units or more	4
$f_s f_s$	Speed of moving system (FM 46-IV)	37
G	Period of time covered by forecast (FM 61-IV)	48, **49**
GG	Time of observation, UTC	4
$G_1 G_1$	Time of commencement of period of forecast (FM 61-IV)	47
$G_c G_c$	Synoptic hour in UTC of observation of data from which chart is prepared (FM 46-IV)	37
$G_p G_p$	Number of hours to be added to $G_c G_c$ (chart time) to obtain time to which the prognosis refers (FM 46-IV)	37
$g_p g_p$	Number of hours to be added to, or subtracted from $G_c G_c$ to give supplementary information (FM 46-IV)	37
$H_w H_w$	Height of wind waves in ½ metres	6, **48**
$H_{w1} H_{w1}$	Height of first swell waves in ½ metres	6
$H_{w2} H_{w2}$	Height of second swell waves in ½ metres	6
$H_{wa} H_{wa}$	Measured height of waves in ½ metres	6
$H_{wa} H_{wa} H_{wa}$	Measured height of waves in tenths of a metre	7
h	Height of base of Low or Medium Cloud	4, **9**
I_s	Type of ice accretion	7, **24**
i_R	Indicator for precipitation group	4, **8**
i_w	Wind speed indicator	4, **8**
i_x	Indicator for weather group	4, **8**
II	Regional block number (**FM 12-X Ext.**)	46
iii	Station number (**FM 12-X Ext.**)	46
k	Indicator for specifying the half degree of latitude and longitude (FM 46-IV)	36, **40**
$L_a L_a$	Latitude to nearest degree (FM 46-IV)	37
$L_o L_o$	Longitude to nearest degree (FM 46-IV)	37

Symbol	Meaning	Page
$L_aL_aL_a$	Latitude in degrees and tenths	4
$L_oL_oL_oL_o$	Longitude in degrees and tenths	4
m	Movement indicator figure (FM 46-IV)	37, **41**
N	Total amount of cloud	4, **45**
N_h	Amount of Low Cloud or Medium if no Low Cloud present	5, **11**
$n_bn_bn_b$	Serial number of offshore platform	4
P_c	Character of pressure system (FM 46-IV)	37, **40**
P_t	Type of pressure system (FM 46-IV)	37, **40**
P_w	Period of waves in code (FM 61-IV)	48, **51**
P_wP_w	Period of sea waves	6, 38
$P_{w1}P_{w1}$	Period of first swell waves in seconds	6
$P_{w2}P_{w2}$	Period of second swell waves in seconds	6
$P_{wa}P_{wa}$	Measured period of waves in seconds	6
PP	Pressure in whole millibars (FM 46-IV)	37
PPP	Pressure in whole millibars of an isobar (FM 46-IV)	38
PPPP	Pressure in millibars and tenths	5
ppp	Amount of barometric tendency	5
Q	Octant of the globe (FM 46-IV)	37, 38, **40**
Q_c	Quadrant of the globe	4, **8**
R_s	Rate of ice accretion	7, **24**
RRR	Amount of precipitation	5
S	State of sea (FM 61-IV)	48, **50**
S_i	Stage of development of sea ice	7, **25**
s_n	Sign of temperature in code	5, **14**
s_s	Indicator for sign and type of measurement of sea surface temperature	6, **26**
s_w	Indicator for the sign and type of wet bulb temperature recorded	7, **26**
T_c	Tropical system characteristic (FM 46-IV)	38, **42**
T_i	Tropical system intensity (FM 46-IV)	38, **42**
T_n	Minimum air temperature (FM 61-IV)	48, **51**
T_t	Tropical circulation type (FM 46-IV)	38, **42**
T_x	Maximum air temperature (FM 61-IV)	48, **51**
TT	Air temperature in whole degrees	3
TTT	Air temperature in whole degrees and tenths	5
$T_bT_bT_b$	Wet bulb temperature in whole degrees and tenths, its sign being given by s_w	7, **26**
$T_dT_dT_d$	Dew-point temperature in whole degrees and tenths	5, **27**
$T_wT_wT_w$	Sea temperature in whole degrees and tenths	6
t_R	Duration of precipitation	5
uu	Isopleth values (FM 46-IV)	38
V	Horizontal visibility (FM 61-IV)	48, **50**
VV	Horizontal visibility	4, **10**
v_s	Ship's speed	6, **24**
W_1W_2	Past weather	5, **21**, **45**
W_m	Forecast weather (FM 61-IV)	48, **50**
ww	Present weather	5, **15**
w_sw_s	Significant weather (FM 46-IV)	38, **41**
YY	Day of the month, UTC	4, 37, 47
z_i	Ice situation and trend of conditions	7, **26**

INDEX

Where reference is to a code table the page number is in **bold type**.

Analysis:
  block numbers, 46
  charts for, 33, 43
  code, 36–46
  example, 33
  explanation, 39
  land station, code for, 46
  map, 60
  specification tables, **40–42**
Atlantic Weather Bulletin:
  analysis, 33
  contents of, 31
  example of, 32
  land station report, 44, 46
  map, 60
  plotting, 34
  ships' reports, 44
  station model, 43
  symbols for, 35
  transmission of, 31
  use of, 31

Barometric:
  characteristic, 5, **14**
  pressure, 5
  tendency, 5
Beaufort:
  letters, **35**
  wind force, 4, **12**, 48, **49**
Block numbers, 46

Cloud:
  amount, 4, 5, **11**, **25**
  form of low, 5, **21**
  form of medium, 5, **22**
  form of high, 5, **23**
  height, 4, **9**
Code forms:
  auxiliary ships, 3
  I.A.C. Fleet, 36
  land stations, 46
  offshore platforms, 3
  MAFOR, 47
  selected ships, 3
  supplementary ships, 3
Coded groups:
  analysis code, 36–39
  land station, 46
  MAFOR, 48–51
  omissions of, 3
  ships' reports, 4–7
Code letters, 56
Confidence group, 38, **42**

Day of month, 4, 37
Dew-point, 5, 14, **27–30**
Direction of surface wind, 4, **11**

Facsimile symbols, 54
Forecasts:
  MAFOR, 47
  North Atlantic, 31–35
  period of time covered by, 47, 48, **49**
Fronts:
  character of, 38, **41**
  intensity of, 38, **41**
  symbols for, **35**
  type of, 38, **41**

Gale:
  areas, 52
  warnings, 32

Half-degree position, 36, **40**
Height of cloud, 4, **9**
High cloud, form of, 5, **23**
I.A.C. FLEET, 36–46
Ice:
  bearing of ice edge, 7, **26**
  concentration or arrangement of, 7, **24**
  of land origin, 7, **25**
  situation and trend, 7, **26**
  stage of development, 7, **25**
Ice accretion:
  rate, 7, **24**
  thickness, 7
  type, 7, **24**
ICING, 7
Isopleth, types of, 38, 39, **42**

Land station reports, 44, 46
Latitude in position groups, 4, 36–39, **40**
Longitude in position groups, 4, 36–39, **40**
Low cloud, type of, 5, **21**

MAFOR code, 48–51
Maritime forecast code, 47
Maritime indicator groups, 47, **48**
Medium cloud, Type of, 5, **22**
Movement:
  group, 36–39, **41**
  indicator figure, 36–39, **41**
  of systems, 34, 36–39

Octant of the globe, 37, **40**

Past weather:
  code letters, 5
  code table, **21**
  plotting symbols, **45**
Period of:
  sea waves, 6, 48
  swell waves, 6, 48
Plotting of observations, 43

INDEX — *continued*

Port Meteorological Officers, 61
Present weather:
   code letters, 5
   code table, **15–20**
   plotting symbols, **45**
Pressure:
   barometric, 5
   character, 37, **40**
   characteristic, 5, **14**
   tendency, 5
   type of, 37, **40**

Quadrant of the globe, 4, **8**

Sea:
   state, 48, **50**
   temperature, 6, **26**
   waves, 6, 38, 48, **50**
Ship's course, 6, **23**
Ships' reports, 44
Significant weather, 38, **41**
State of sea, 48, **50**
Station numbers, 46
Station reports, 44, 46
Swell, 6, **11**
Symbols used for plotting:
   cloud amount, **45**
   fronts, **35**
   past weather, **45**
   present weather, **45**
   wind direction and speed, **45**

Temperature:
   air, 5
   dew-point, 5, 14, **27–30**
   maximum, 48, **51**
   minimum, 48, **51**
   sea, 6
   wet bulb, 7, **26**
Time of observation, 4
Total amount of cloud, 4, **11**, **45**
Tropical:
   circulation type, 38, **42**
   system characteristic, 38, **42**
   system intensity, 38, **42**

Visibility at the surface:
   MAFOR code, 48, **50**
   Ships' code 4, **10**

Waves:
   direction, 6, **11**, 38
   height of, 6, 7, **53**
   period of, 6, 38
Weather Bulletins:
   map, 52
Wind:
   Beaufort force, 4, **12**, 48, **49**
   direction, 4, **11**, 48, **49**
   metres per second, 45
   plotting symbols, **45**
   speed, 4, **11**, 48, **49**

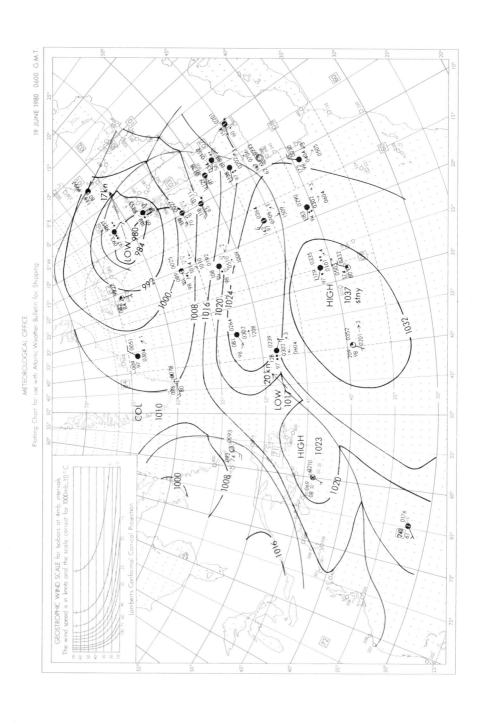

PORT METEOROLOGICAL OFFICES

South-east England — PMO, Trident House, 21 Berth, Tilbury Dock, Tilbury, Essex RM18 7HL. Telephone: 01375 859970. Fax: 01375 859972.

North-west England — PMO, Room 331, Royal Liver Building, Liverpool L3 1JH. Telephone: 0151 236 6565. Fax: 0151 227 4762.

Scotland and Northern Ireland — PMO, Navy Buildings, Eldon Street, Greenock, Strathclyde PA16 7SL. Telephone: 01475 724700. Fax: 01475 892879.

South-west England — PMO, 8 Viceroy House, Mountbatten Business Centre, Millbrook Road East, Southampton SO15 1HY. Telephone: 01703 220632. Fax: 01703 337341.

Bristol Channel — PMO, P.O. Box 278, Companies House, Crown Way, Cardiff CF4 3UZ. Telephone: 01222 221423. Fax: 01222 225295.

East England — PMO, Customs Building, Albert Dock, Hull HU1 2DP. Telephone: 01482 320158. Fax: 01482 328957.

North-east England — PMO, Able House, Billingham Reach Industrial Estate, Billingham, Cleveland TS23 1PX. Telephone: 01642 560993. Fax: 01642 562170.